THE STAR & CROSS POLYHEDRA

BEING THE FOURTH PART
OF SEVERAL COMPRISING
THE COMPLETE? POLYHEDRA
BY PATRICK TAYLOR

THE STAR & CROSS POLYHEDRA

First published 2000 by

Nattygrafix
277 Cavendish Street
Ipswich
Suffolk
IP3 8BQ

(01473) 289861

Made in England

ISBN 0 9516701 5 8

NATTYGRAFIX IPSWICH

CONTENTS

ILLUSTRATIONS

Figures * are supported by Tables nos. 1-37

for Keith Critchlow, his encouragement and wisdom

"Beware the Jabberwock, my son!
  The jaws that bite, the claws that catch!
Beware the Jubjub bird, and shun
  The frumious Bandersnatch!"

Lewis Carroll *Through the Looking-Glass*
1871

INTRODUCTION

Here be monsters is the traditional wisdom when dealing with the star polyhedra! They seem to sort of fit the pattern of the other polyhedra, but don't quite totally behave. Maybe we have been using the wrong rules?

In the previous volume, '*The Simpler? Polyhedra*', we have already encountered some polyhedra with star and cross polygon faces, both as quasitruncations and as members of various sub-families of the Platonic solids. Many of these could also be seen as possible members of more complex symmetry families, i.e. {4,4/2}, {5,5/2} or {3,5/2}.

This volume, the fourth in the series, will now explore these more complex super-families of the star and cross polyhedra derived from using star and cross polygons both as faces and as vertex figures. In the same way that the last volume paralleled the first part of the second volume, '*Incomplete Tilings*', this one will now parallel the second part of the same.

We will therefore start by looking at two super-families generated from pairs of sub-family Schwarz triangles being combined. We will then continue with the star and cross polyhedra proper and in turn their own sub-families.

As before we will continue to allow the inclusion of compound and double polygons, and consequently paint a more complete? picture of those symmetry families than is generally available. The increasing numbers of curious gaps, where certain truncations are deemed not to exist in the standard classification, will continue to be filled, and hopefully our monsters will be laid to rest.

TWO MORE SCHWARZ TRIANGLES

If we repeat the procedure we followed in '*Incomplete Tilings*' and take the two sub-family Schwarz triangles as components of a larger assembly of four Möbius triangles, we arrive at two new Schwarz triangles in our spherical kaleidoscopes.

These are (3 4/3 4) and (3 5/3 5) as shown in Figure 1, the generating triangles to two new super-families of polyhedra.

The first of these has already been seen in the last volume as a colunar triangle to the (4 3/2 4) sub-family. We will thus not illustrate its members here, but we will include them in our tabulations for ease of comparison and completeness.

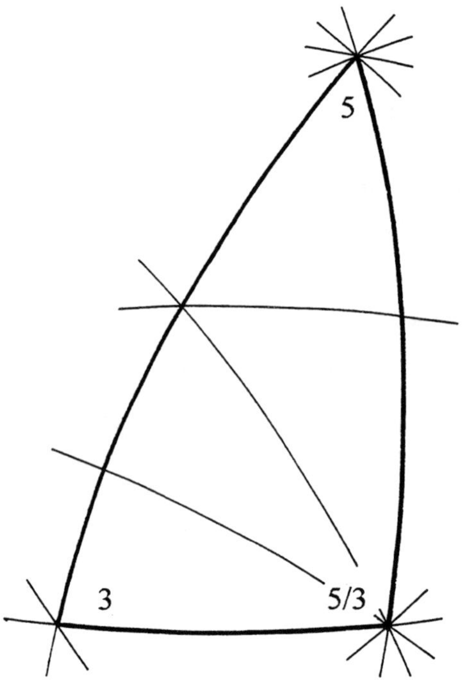

(3 5/3 5)

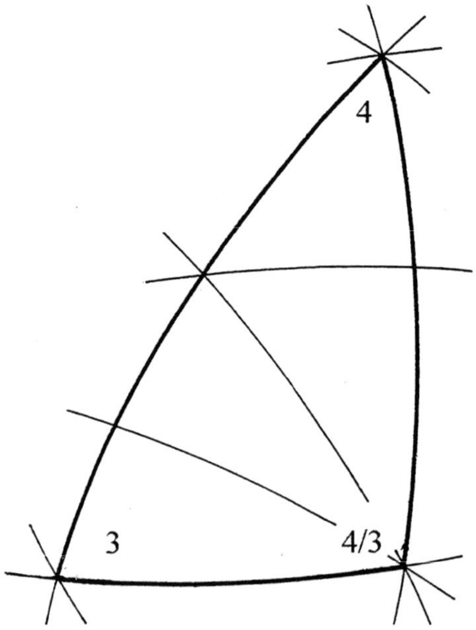

(3 4/3 4)

Fig.1: Two Schwarz Triangles

Symbol	V	E	F
3| 5/3 5 53 (80) Ditrigonal Dodecadodecahedron	20	60	12 ⬠ 12 ☆
5| 3 5/3 Inscribed Small Stellated Dodecahedron	12	30x2	12 ☆ 20 △
5/3| 3 5 Inscribed Icosahedron	12	30x2	12 ⬠ 20 △
3| 4/3 4 17 (2) Double Cube	8	12x2	6 □ 6 □
4| 3 4/3 4/3| 3 4 Inscribed Octahedron	6	12x2	6 □ 8 △

Table 1: Quasi-Regular Polyhedra

SIX QUASI-REGULAR POLYHEDRA

Taking the vertices of these new generating triangles as our starting points, we seem to have potentially six new quasi-regular polyhedra to investigate.

However, two of these, 4| 3 4/3 and 4/3| 3 4, are one and the same polyhedron, so that only five are described opposite in Table 1. Those with fourfold symmetry, the Double Cube and Inscribed Octahedron, appeared in '*The Simpler? Polyhedra*' pp. 45 and 47, whilst the three with fivefold symmetry are shown below in Figure 2.

Two of these are compound polyhedra, whilst the third, 3| 5/3 5, has not been seen before, despite being inscribed within the shell of the ubiquitous Small Ditrigonal Icosidodecahedron.

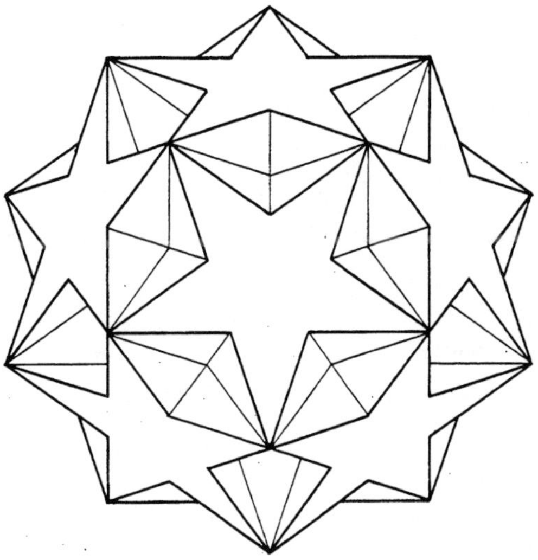

3| 5/3 5
53 (80)
Ditrigonal
Dodecadodecahedron

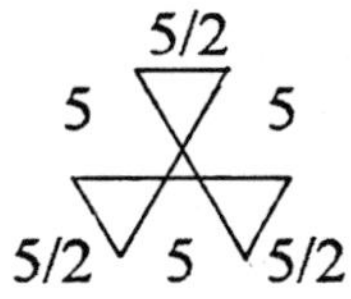

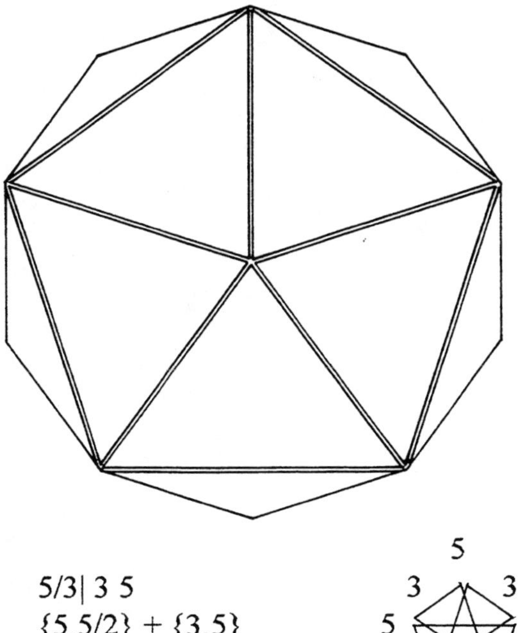

5/3| 3 5
{5,5/2} + {3,5}

Inscribed Icosahedron

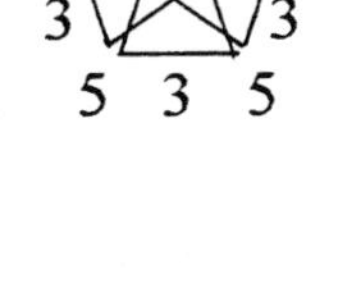

5| 3 5/3
{3,5/2} + {5/2,5}
Inscribed Small
Stellated Dodecahedron

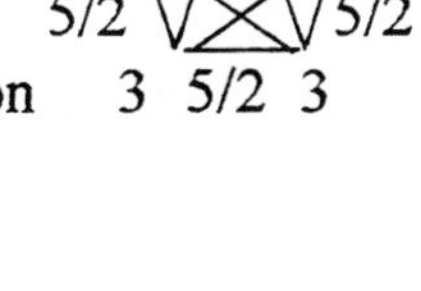

Fig.2: Quasi-Regular Polyhedra

EIGHT EVEN FACED POLYHEDRA

Moving on to the centres of our new Schwarz triangles and their outlying colunar triangles, we can find four even faced polyhedra for each super-family.

Within the fourfold super-family, not illustrated, we again find that two of these are one and the same, so that a total of seven are described below in Table 2. The four even faced members of the fivefold super-family are shown in Figure 3.

Symbol	V	E	F
3 5/3 5|	120	180	12
57 (84)			12
Icosidodecatr. Icosidodecahedron			20
3 5/2 5/4|	30x4	60x3	12
			12
Inscribed Dodecadodecahedron			20
3/2 5/2 5|	30x4	60x3	12
			12
Inscribed Icosidodecahedron			20
3/2 5/3 5/4|	30x4	60x3	12
			12
Inscribed Great Icosidodecahedron			20
3 4/3 4| = 3 4 4/3|	48	72	6
52 (79)			6
Cuboctatruncated Cuboctahedron			8
3/2 4 4|	24x2	36x2	6
			6
Double Truncated Cube			8
3/2 4/3 4/3|	24x2	36x2	6
			6
Double Quasitruncated Cube			8

Table 2: Even Faced Polyhedra

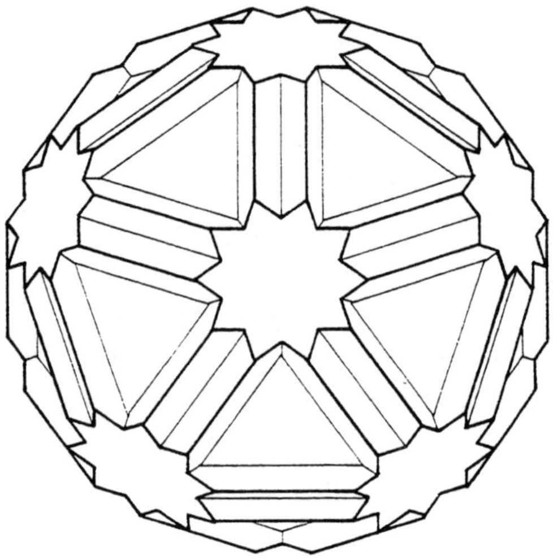

3 5/3 5|

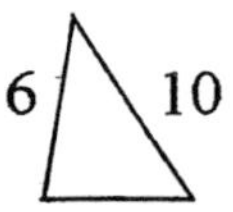

57 (84)

Icosidodecatruncated Icosidodecahedron

3 5/2 5/4|

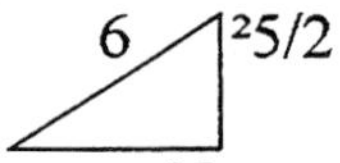

Inscribed Dodecadodecahedron

Fig.3: Even Faced Polyhedra

SIX RHOMBIHEDRA

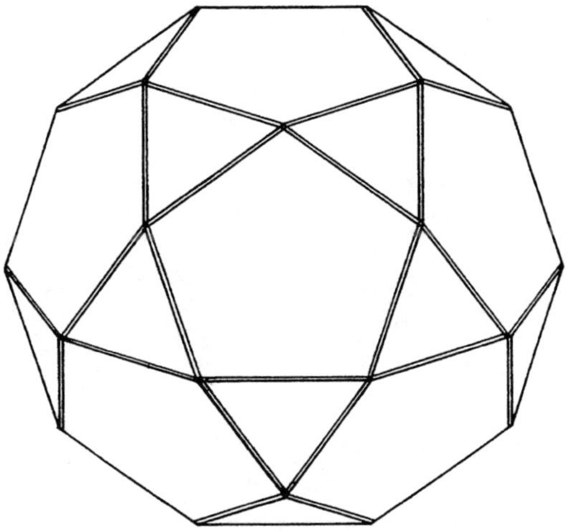

If we next consider the edges of our new Schwarz triangles, we can find the vertices of three rhombihedra in each super-family.

All six are described below in Table 3 and those with fivefold symmetry shown overleaf in Figure 4.

Whilst the fourfold rhombihedra have been seen before in '*The Simpler? Polyhedra*', pp. 49, 53 and 55, the fivefold ones have not been encountered in our journey so far, and will in due course produce some further p-type and q-type rhombihedra.

3/2 5/2 5|

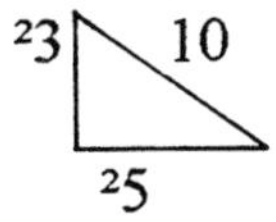

Inscribed Icosidodecahedron

3/2 5/3 5/4|

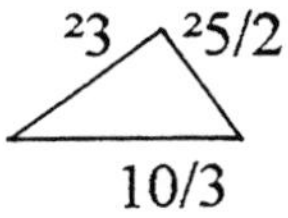

Inscribed Great Icosidodecahedron

Symbol	V	E	F
3 5 |5/3	60	120	12 ⬠
54 (81)			12 ✺
Gt. Ditr. Dodecicosidodecahedron			20 △
3 5/3 |5	60	120	12 ○
55 (82)			12 ☆
Sm. Ditr. Dodecicosidodecahedron			20 △
5 5/3 |3 p{5, 5/2}	60	120	12 ⬠
56 (83)			12 ☆
Icosidodecadodecahedron			20 ⬡
3 4 |4/3	24	48	6 □
50 (77)			6 ✷
Great Cubicuboctahedron			8 △
3 4/3 |4 (= 3/2 4 |4)	24	48	6 ○
38 (69)			6 □
Small Cubicuboctahedron			8 △
4 4/3 |3	12x2	24x2	6 □
51 (78)			6 □
Cubohemioctahedron			8 ⬡

Table 3: Rhombihedra

3 5 |5/3
54 (81)
Great Ditrigonal
Dodecicosidodecahedron

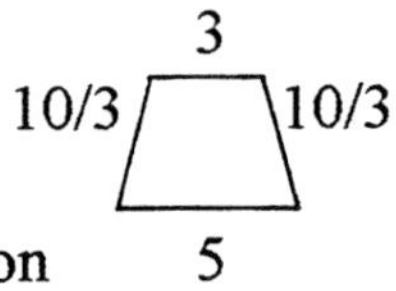

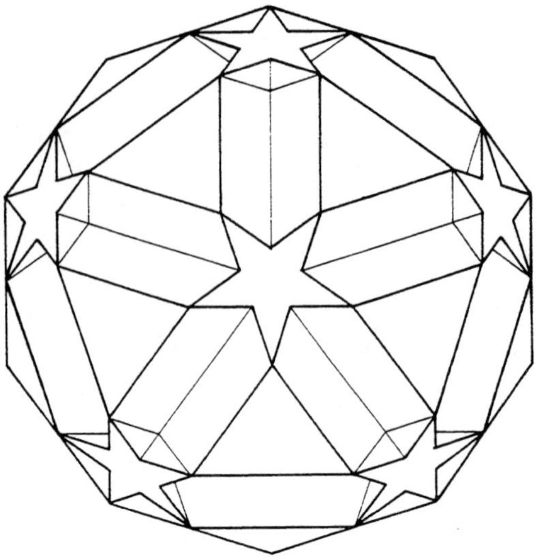

3 5/3 |5
55 (82)
Small Ditrigonal
Dodecicosidodecahedron

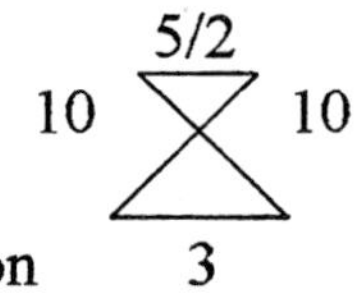

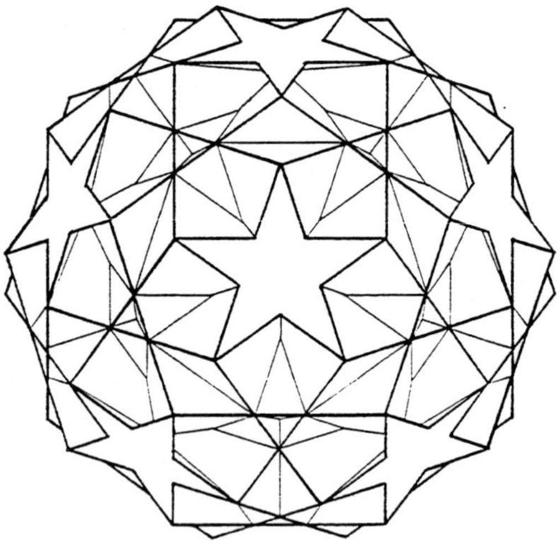

5 5/3 |3 p{5, 5/2}
56 (83)
Icosidodecadodecahedron

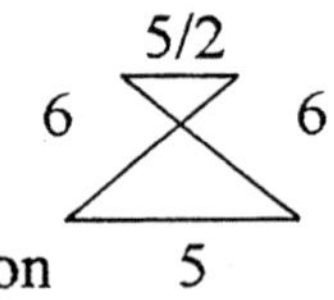

Fig.4: Rhombihedra

Symbol	V	E	F
3/2 5 |5/2 Inscribed Icosahedron	12x5	30x4	12 ⬠ 12 ⬠ 20 △
3 5/2 |5/4 Inscribed Small Stellated Dodecahedron	12x5	30x4	12 ☆ 12 ☆ 20 △
5 5/2 |3/2 p'{5, 5/2} Quasicosidodecadodecahedron	20x3	60x2	12 ⬠ 12 ☆ 20 ▲
3/2 4 |4 3 4 |4/3	see Table 3		
4 4 |3/2 Double Cuboctahedron	12x2	24x2	6 □ 6 □ 8 ▲

Table 4: Quasirhombihedra

SIX QUASIRHOMBIHEDRA

Not surprisingly, the extended edges of our new Schwarz triangles also yield polyhedra, this time three quasirhombihedra for each super-family.

The three with fivefold symmetry are shown below in Figure 5, and all are described opposite in Table 4, except for two of the fourfold super-family, 3/2 4 |4 and 3 4 |4/3, which are repeats of those already described in Table 3 on the previous page.

The inscribed fivefold forms shown here all have multiple vertices and multiple edges as shown in Table 4. These result from the double polygon faces that occur in their vertex figures, as will be explained later on p.27.

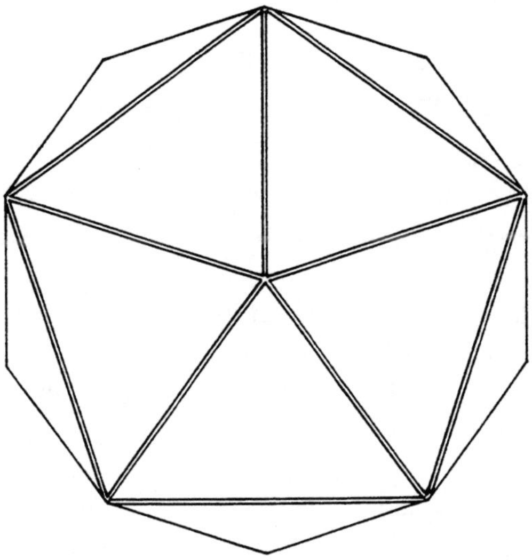

3/2 5 |5/2

{3,5} + 3{5,5/2}
Inscribed Icosahedron

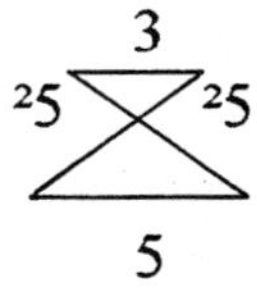

5 5/2 |3/2 $p'\left\{\begin{matrix}5\\5/2\end{matrix}\right\}$

Quasicosidodecadodecahedron

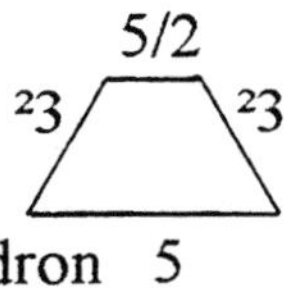

3 5/2 |5/4
{3,5/2} + 3{5/2,5}
Inscribed Small
Stellated Dodecahedron

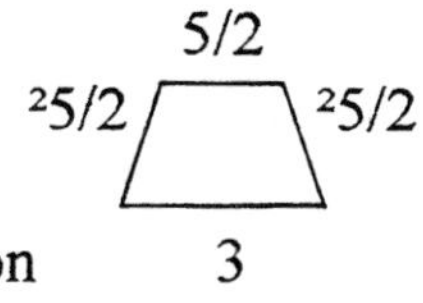

Fig.5: Quasirhombihedra

FOUR SNUB POLYHEDRA

Following on from examining the rhombihedra and quasirhombihedra it is natural enough to ask whether our new super-families yield any corresponding snub polyhedra. As might by now be expected they produce a snub and quasisnub for each super-family as described below in Table 5.

Again those with fourfold symmetry have already been described in the previous volume, so that Figure 6 shows only the two versions with fivefold symmetry.

Because these (3 n/3 n) super-families have four Möbius triangles in each Schwarz triangle, they have as expected fourfold density. This is perhaps most easily seen in the Inscribed Icosahedron (p.7), where any single triangle is undercut by three different pentagons between the outside and the centre of the polyhedron.

Symbol	V	E	F
|3 5/3 5 58 (112) Snub Icosidodecadodecahedron	60	180	12 ⬠ 12 ✯ 80 △
|3/2 5/2 5 Quasisnub Icosidodecadodecahedron	20x3	60x3	12 ⬠ 12 ✯ 80 △
|3 4/3 4 36 (67) Double Tetrahemihexahedron	6x4	12x6	6 □ 6 □ 32 △
|3/2 4 4 19 (11) Cuboctahedron	12x2	24x3	6 □ 6 □ 32 △

Table 5: Snub Polyhedra

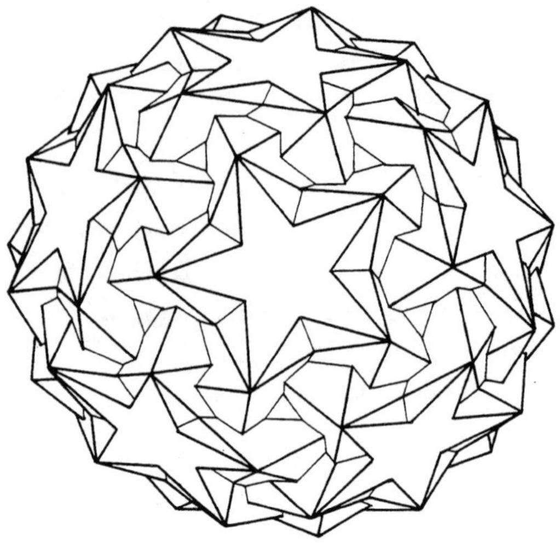

|3 5/3 5
58 (112)
Snub
Icosidodecadodecahedron

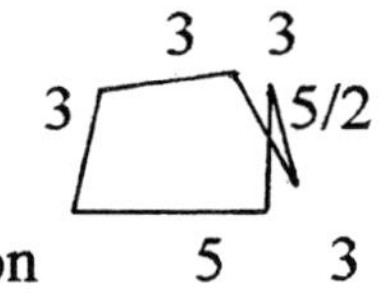

|3/2 5/2 5
Quasisnub
Icosidodecadodecahedron

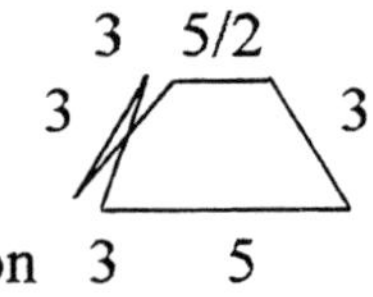

Fig.6: Snub Polyhedra

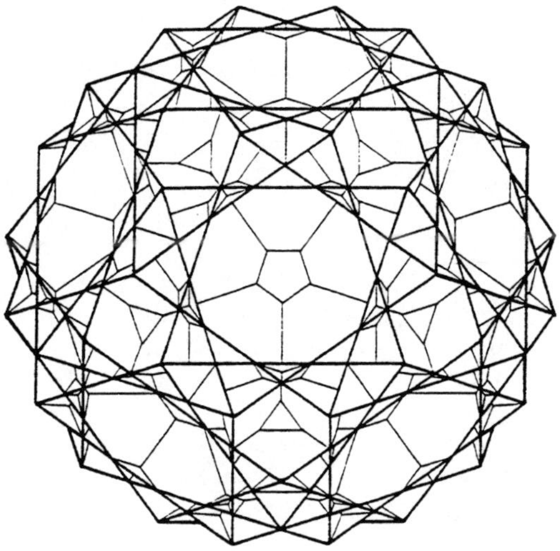

3/2 5 |3

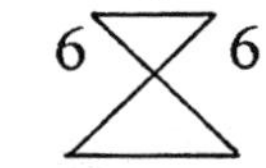

62 (88)
Great Icosicosidodecahedron

3 5/3 3/2|
5/2|

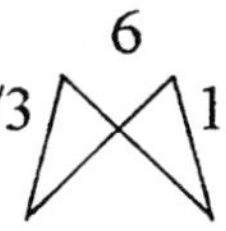

79 (101)
Great Dodecicosahedron

Fig.7: P- and Q-type Rhombihedra

P- AND Q-TYPE RHOMBIHEDRA

The rhombihedra we have just examined, like those in the last volume, will also have p- and q-type rhombihedra associated with them, formed by taking the overall six faces in the vertex figure and missing pairs out.

Those from the fourfold symmetry super-family were considered in the last volume, whilst the fivefold quasirhombihedra, illustrated in Figure 5, produce nothing new other than further inscribed forms. However the fivefold rhombihedra are more fruitful and produce six more polyhedra.
Two of these, derived from 3 5/3 |5, were considered in the last volume (p.57) and a further two, derived from 5 5/3 |3, will reappear later in this one, so that here in Figure 7 we only show the other two, derived from the 3 5 |5/3 rhombihedron.

TWO SUPER-FAMILIES

Because of their fourfold density, the members of these (3 n/3 n) super-families, have a corresponding Euler's Formula, V+F=E+8 (since 8 is twice the density). This can be seen to work with the figures tabulated if we count three times all the faces derived from the {n/3}.

The Schwarz triangles and associated colunar triangles of the two super-families are illustrated overleaf in Figures 8 and 9. Each has produced three quasi-regular polyhedra at the vertices of the Schwarz triangle (3| n/3 n, n| 3 n/3 and n/3| 3 n), four even faced polyhedra at the centres of the colunar triangles (3 n/3 n|, 3 n/3' n'|, 3/2 n/3' n| and 3/2 n/3 n'|), three rhombihedra (3 n |n/3, 3 n/3 |n and n n/3 |3), three quasirhombihedra (3/2 n |n/3', 3 n/3' |n' and n n/3' |3/2), a snub (|3 n/3 n) and a quasisnub (|3/2 n/3' n) polyhedron.

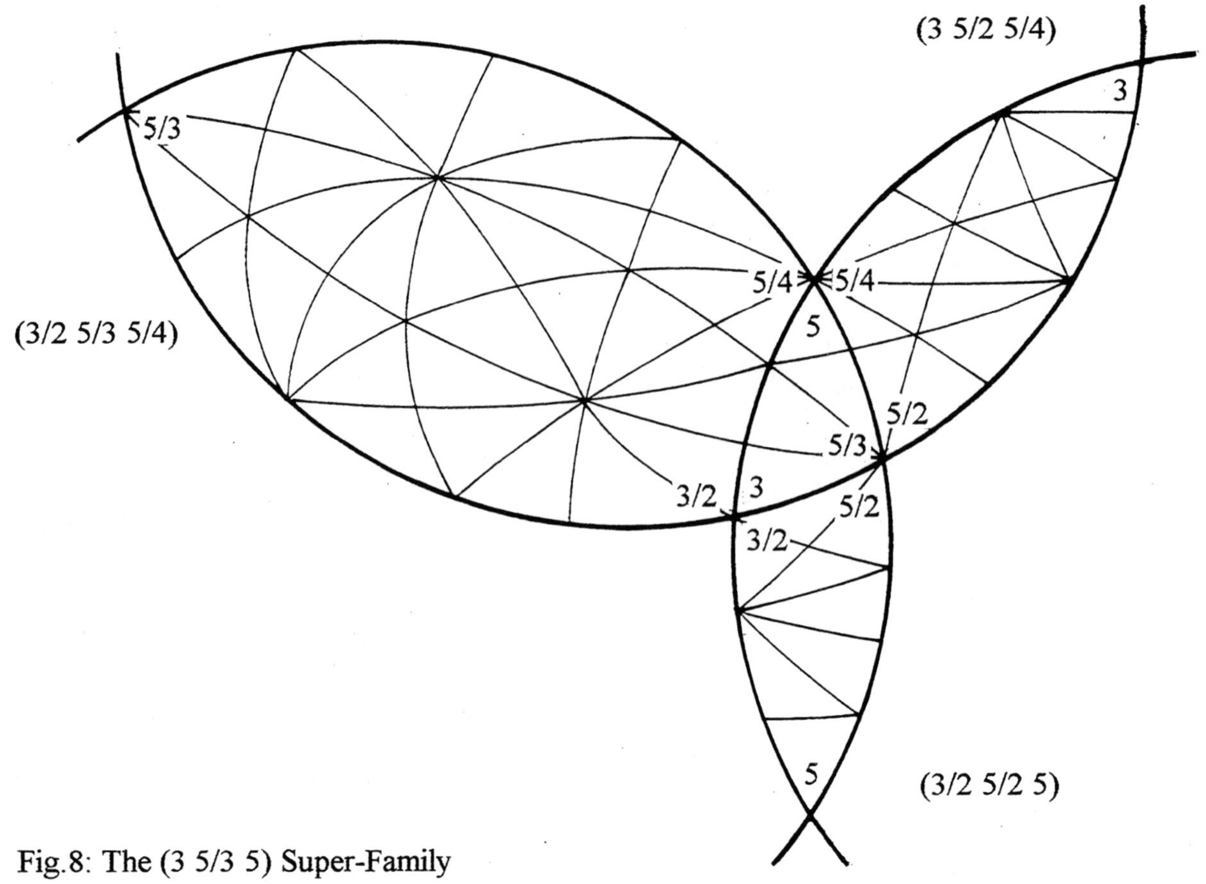

Fig.8: The (3 5/3 5) Super-Family

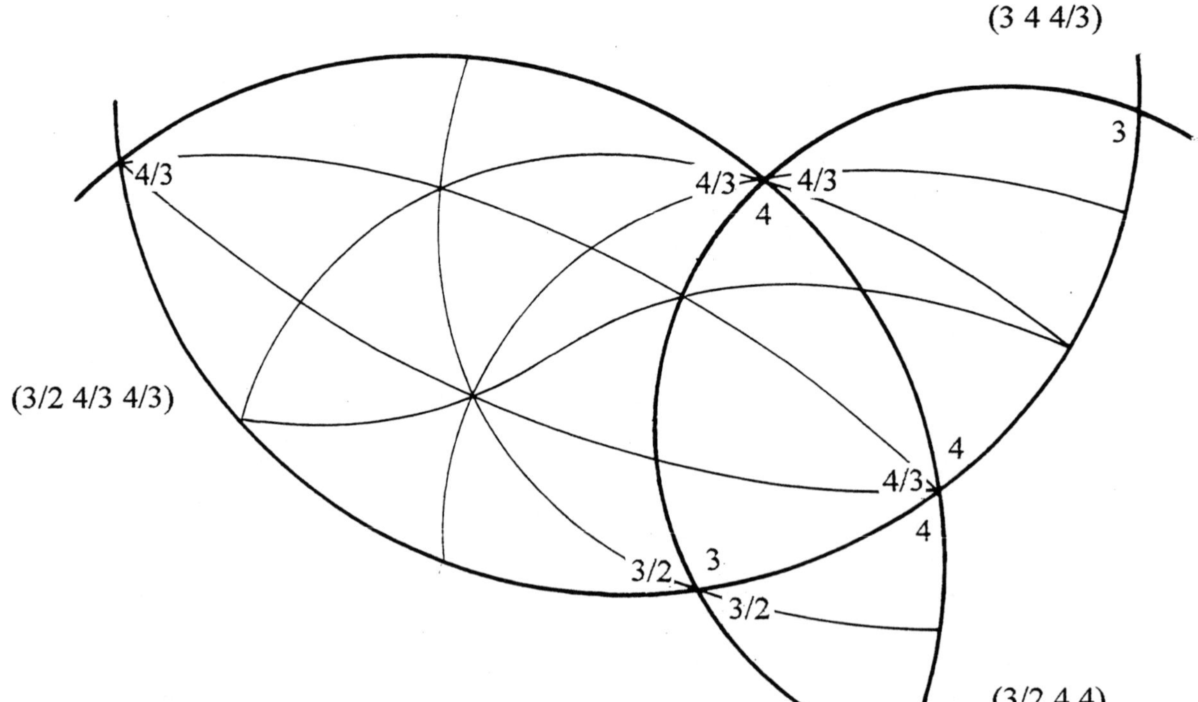

Fig.9: The (3 4/3 4) Super-Family

THREE STAR FAMILIES

Throughout the last volume, '*The Simpler? Polyhedra*', the polyhedra derived from the {3,3}, {3,4} and {3,5} symmetry families gave us a great many hints that there might be further symmetry families to examine, namely {4,4/2}, {5,5/2} and {3,5/2}.

Looking at our spherical kaleidoscopes, we can identify Schwarz triangles corresponding to these i.e. (4 2 4/2), (5 2 5/2) and (3 2 5/2) comprised of 3, 3 and 7 Möbius triangles respectively.

These are shown in Figure 10 and are the generating triangles for one family of cross polyhedra, with cross polygon {4/2} faces, and two families of star polyhedra, with star polygon {5/2} faces.

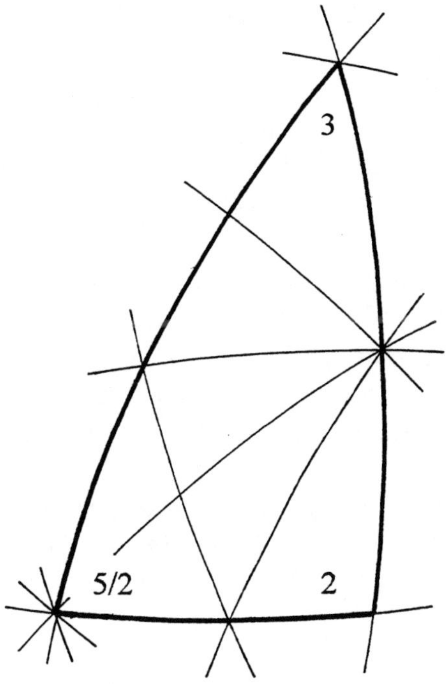

(3 2 5/2)

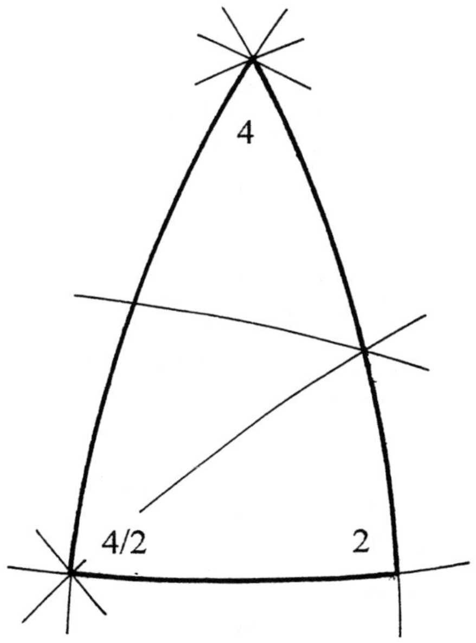

(4 2 4/2)

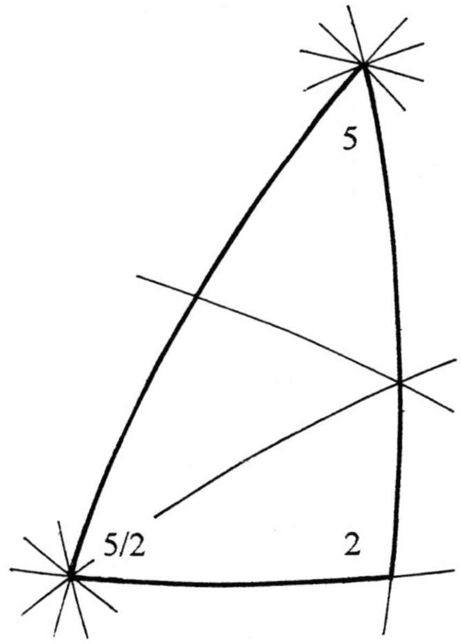

(5 2 5/2)

Fig. 10: Cross & Star Family Schwarz Triangles

REGULAR STAR POLYHEDRA

Taking the vertices of these new Schwarz triangles (other than {2}) as the generating points for new polyhedra we can obtain two regular polyhedra for each of these families.

These are shown in Figures 11 & 12 and comprise two cross polyhedra in the fourfold family and four star polyhedra (known as the Kepler-Poinsot solids after their discoverers) in the fivefold families. The vertices, edges and faces of these are listed below in Table 6, and can be seen to form dual pairs, as did the regular polyhedra, where face and vertex symmetries are reversed.

Symbol	V	E	F
4/2| 2 4 {4,4/2} [1.3a] Great Hexahedron	6	12	6 □
4| 2 4/2 {4/2,4} [1.3d] Stellated Hexahedron	6	12	6 ✳
5/2| 2 5 {5,5/2} 44 (21) Great Dodecahedron	12	30	12 ⬠
5| 2 5/2 {5/2,5} 43 (20) Small Stellated Dodecahedron	12	30	12 ☆
5/2| 2 3 {3,5/2} 69 (41) Great Icosahedron	12	30	20 △
3| 2 5/2 {5/2,3} 68 (22) Great Stellated Dodecahedron	20	30	12 ☆

Table 6: Regular Cross & Star Polyhedra

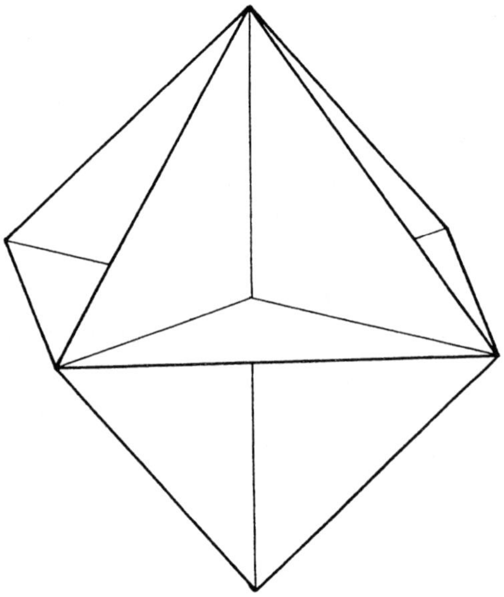

4/2| 2 4 {4,4/2}

[1.3a]
Great Hexahedron

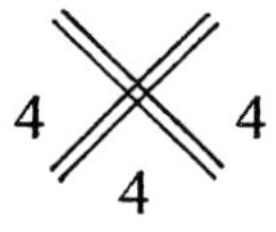

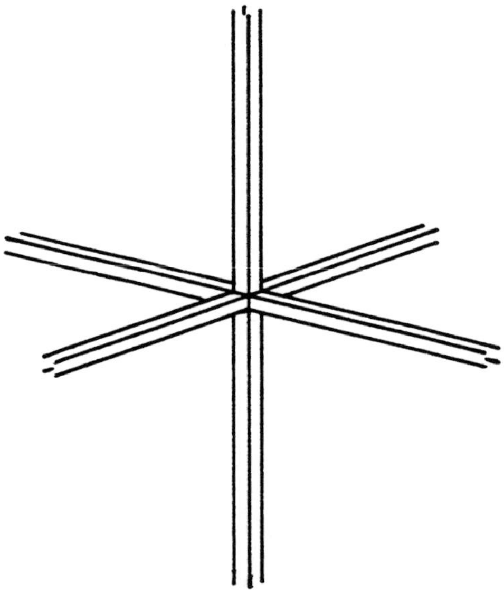

4| 2 4/2 {4/2,4}

[1.3d]
Stellated Hexahedron

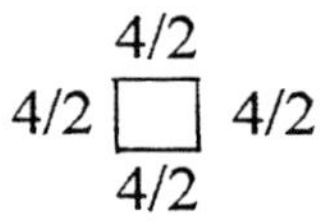

Fig.11: Regular Cross Polyhedra

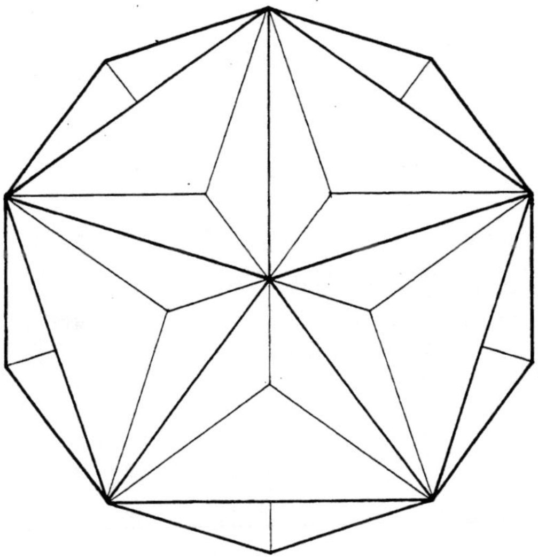

5/2| 2 5 {5,5/2}

44 (21)
Great Dodecahedron

5/2| 2 3 {3,5/2}

69 (41)
Great Icosahedron

5| 2 5/2 {5/2,5}

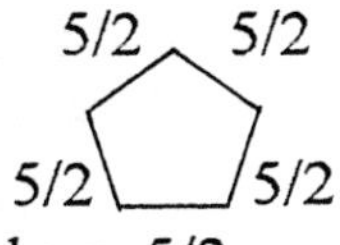

43 (20)
Small Stellated Dodecahedron 5/2

3| 2 5/2 {5/2,3}

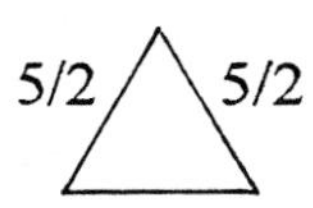

68 (22)
Great Stellated Dodecahedron 5/2

Fig.12: Regular Star Polyhedra (Kepler-Poinsot Solids)

TRUNCATED STAR POLYHEDRA

Whilst three of these new regular cross and star polyhedra are stellations of the cube or dodecahedron, the other three can be seen to be inscriptions within their duals, the octahedron and icosahedron. Taking each of these in turn, we can truncate the vertices to produce semi-regular truncated cross and star polyhedra, as shown in Figures 13 & 14.

These all have the usual isosceles triangle vertex figures, consisting of two truncated polygons from the original faces and one new truncating polygon equivalent to the original vertex figure, all as described below in Table 7.

Symbol		V	E	F
4/2 2 |4 [1.3b] Truncated Great Hexahedron	t{4,4/2}	24	36	6 6
4 2 |4/2 [1.3e] 3{4,3} Truncated Stellated Hexahedron	t{4/2,4}	8x3	12x3	6 6
5/2 2 |5 47 (75) Truncated Great Dodecahedron	t{5,5/2}	60	90	12 12
5 2 |5/2 3{5,3} Truncated Sm Stellated Dodecahedron	t{5/2,5}	20x3	30x3	12 12
5/2 2 |3 71 (95) Truncated Great Icosahedron	t{3,5/2}	60	90	20 12
3 2 |5/2 {3,5} + 2{5,5/2} Truncated Gt Stellated Dodecahedron	t{3,5/2}	12x5	30x3	12 20

Table 7: Truncated Cross & Star Polyhedra

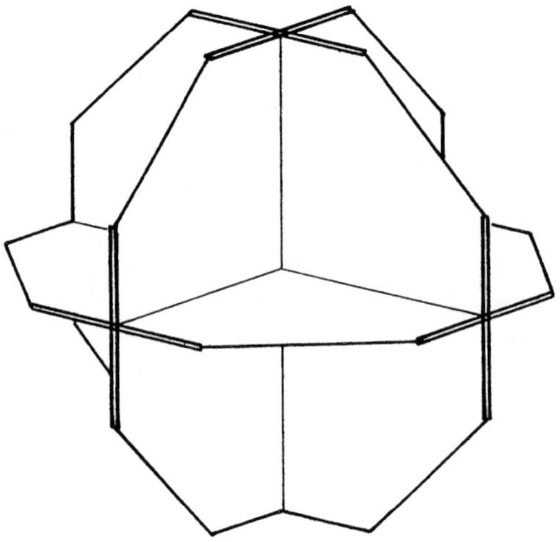

4/2 2 |4 t{4,4/2}

[1.3b]
Truncated Great Hexahedron 4/2

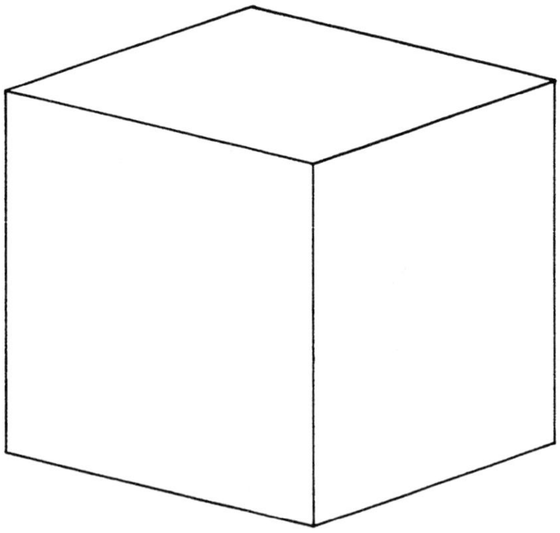

4 2 |4/2 t{4/2,4}
[1.3e] 3{4,3}
Truncated
Stellated Hexahedron

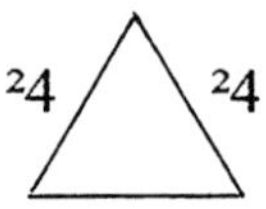

Fig.13: Truncated Cross Polyhedra

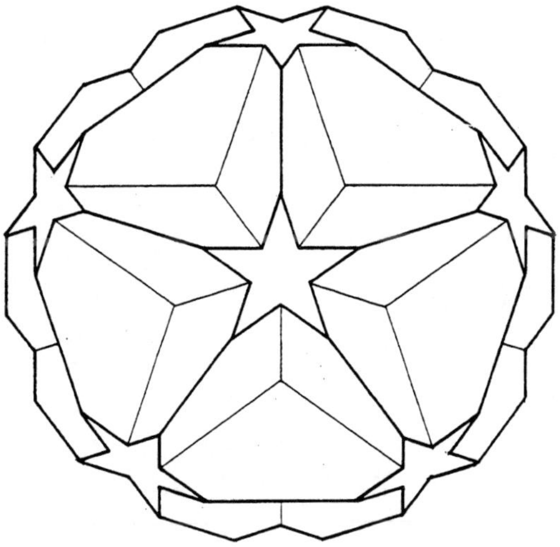

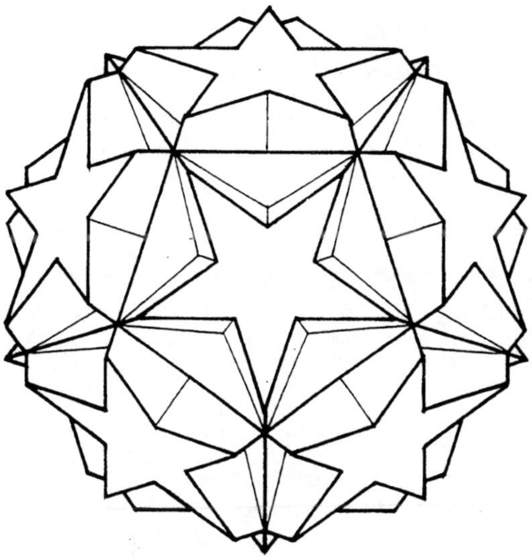

5/2 2 |5 t{5,5/2}

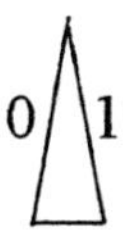

47 (75)
Truncated Great Dodecahedron

5/2 2 |3 t{3,5/2}

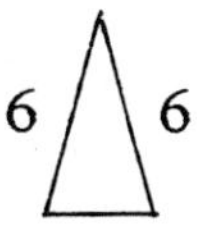

71 (95)
Truncated Great Icosahedron

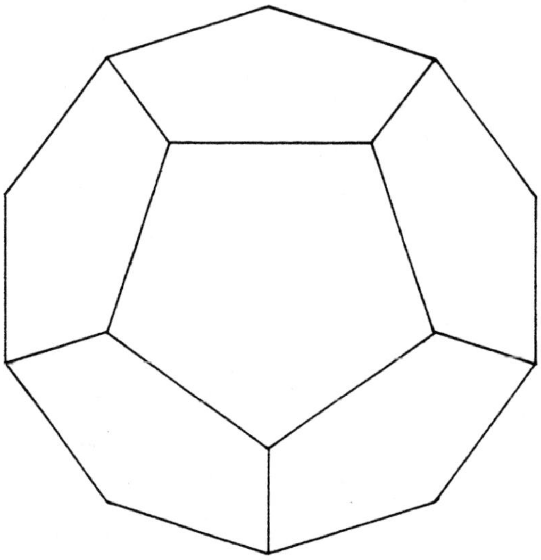

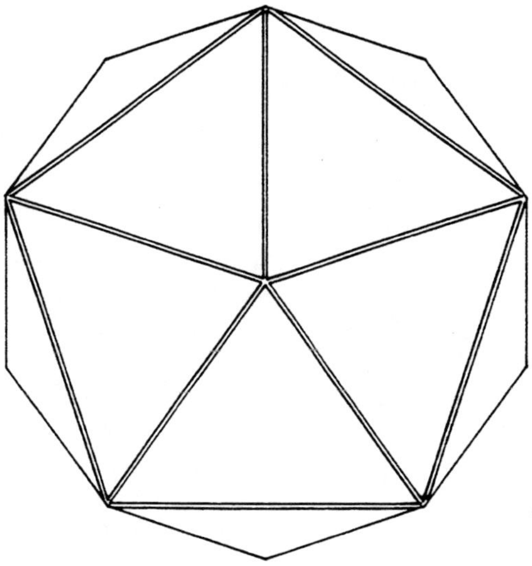

5 2 |5/2 t{5/2,5}
3{5,3}
Truncated Small
Stellated Dodecahedron

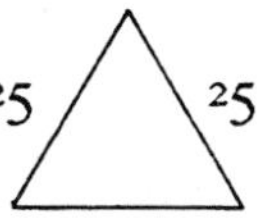

3 2 |5/2 t{5/2,3}
{3,5} + 2{5,5/2}
Truncated Great
Stellated Dodecahedron

Fig. 14: Truncated Star Polyhedra

QUASI-REGULAR STAR POLYHEDRA

As with the regular polyhedra, further truncation of either truncated form leads on to the same central figure in a truncation series, a quasi-regular cross or star polyhedron for each new symmetry family.

This is in fact the polyhedron produced at the {2} vertex of the Schwarz triangle and is shown for each symmetry family below in Figure 15 and described in Table 8 opposite.

As before we get rectangular vertex figures comprised of opposed pairs of the parent polygons. With the inclusion of this quasi-regular polyhedron each symmetry family now has a complete first truncation series.

2| 3 5/2 $\begin{Bmatrix} 3 \\ 5/2 \end{Bmatrix}$

70 (94)

Great Icosidodecahedron

Vertex figure: 5/2, 3, 3, 5/2

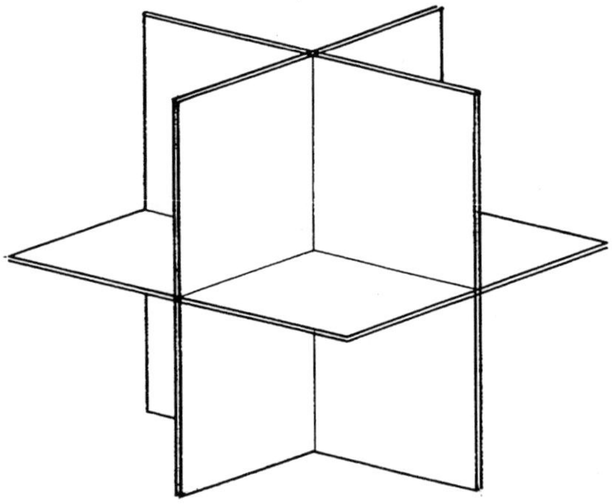

2| 4 4/2 $\begin{Bmatrix} 4 \\ 4/2 \end{Bmatrix}$

[1.3f]

Hexahexahedron

Vertex figure: 4/2, 4, 4, 4/2

2| 5 5/2 $\begin{Bmatrix} 5 \\ 5/2 \end{Bmatrix}$

45 (73)

Dodecadodecahedron

Vertex figure: 5/2, 5, 5, 5/2

Fig.15: Quasi-Regular Cross & Star Polyhedra

EULER'S FORMULA & DENSITY

As we saw in the previous volume with the double density sub-family polyhedra, and at the beginning of this volume with the fourfold density super-families, there is a generalised version of Euler's Formula that appears to work for all polyhedra.

The basic formula for simple, density one, polyhedra is V+F = E+2; however for the more complex polyhedra this is revised to take account of the density of vertices, faces and edges, counting each the requisite number of times and then multiplying the '2' by the density of the polyhedron as a whole.

This overall density is given by the number of Möbius triangles in the generating Schwarz triangle for the family. It can be equated to the number of planes that have to be crossed to get from the outside to the centre of the polyhedron, much as the density of a polygon is a measure of the number of edges that have to be crossed (see *The Complete? Polygon* pp. 34-35).

Looking at the density of our new cross and star family polyhedra, we find for the two simpler families, (4 2 4/2) and (5 2 5/2) with density three, that Euler's Formula can be suitably adjusted by doubling the number of faces derived from the {n/2} and then using V+F = E+6, six being twice the density of three.

However at first sight the (3 2 5/2) family appears not to need such adjustment, as V+F = E+2 seems to work. In general the {5/2} faces in this family number twelve which if included twice as before gives us V+F = E+14, which is of course fine for these polyhedra with density seven.

The way that the density of these polyhedra is built up can be seen by examining the spherical kaleidoscope of Möbius triangles. For instance each Möbius triangle in the (5 2 5/2) family occurs three times, in different positions within three different Schwarz triangles as shown in Figure 16.

Symbol		V	E	F
2| 4 4/2	{ 4 }	12	24	6 □
[1.3f]	{4/2}			6 ✳
Hexahexahedron				
2| 5 5/2	{ 5 }	30	60	12 ⬠
45 (73)	{5/2}			12 ☆
Dodecadodecahedron				
2| 3 5/2	{ 3 }	30	60	20 △
70 (94)	{5/2}			12 ☆
Great Icosidodecahedron				

Table 8: Quasi-Regular Cross & Star Polyhedra

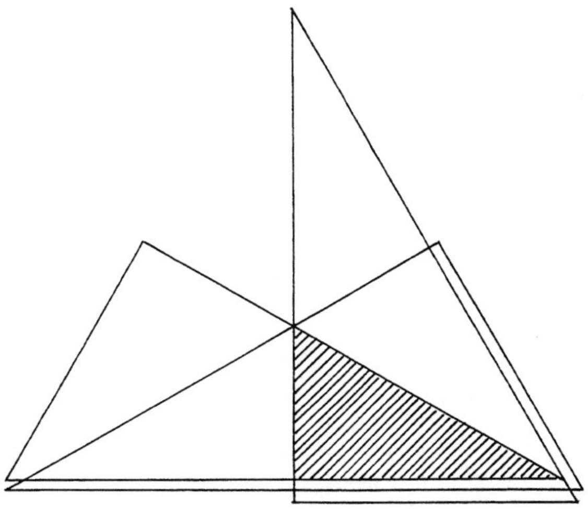

Fig.16: 3-fold Density of Möbius Triangles

RHOMBIC STAR POLYHEDRA

Having found a first series of truncations for each family of cross or star polyhedra we now need to look for a second series. As before we start with the quasi-regular polyhedra and investigate their duals, as formed by simply truncating the vertices to produce at first a new rectangular truncating face, which gradually grows and merges with its fellows to become a rhombic face.

These rhombic polyhedra have two types of vertices and are shown below in Figure 17. The rhombic faces have diagonal lengths proportional to the dual edge lengths of the two polygons comprising the quasi-regular polyhedron. For the (4 2 4/2) and (5 2 5/2) families the rhombs are more pointed so that the (n/2)-fold meeting of faces is hidden within the (n)-fold.

Great Rhombic Triacontahedron

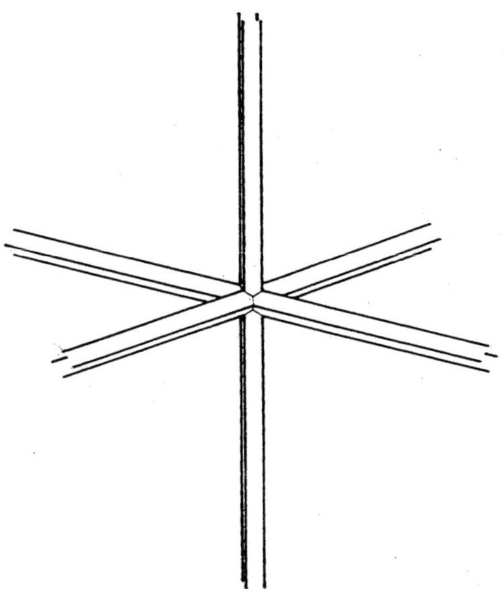

Great Rhombic Dodecahedron

Medial Rhombic Triacontahedron

Fig.17: Rhombic Cross & Star Polyhedra

EVEN FACED STAR POLYHEDRA

As with the regular polyhedra, the first member in a second series starting with the quasi-regular star polyhedra can be achieved by normalising the rectangular faces produced by true truncation into squares.

Since one of the starting faces is always {n/2} this first 'truncation' of a quasi-regular will consist of the truncated faces of the two parent polyhedra, {2x} and $\{^2n\}$, along with a number of squares {4} formed from the rectangles.

These polygons, including the doubles, are all even faced and thus form triangular vertex figures as shown in Figure 18. These even faced cross and star polyhedra represent the centres of our new Schwarz triangles and are described in Table 9.

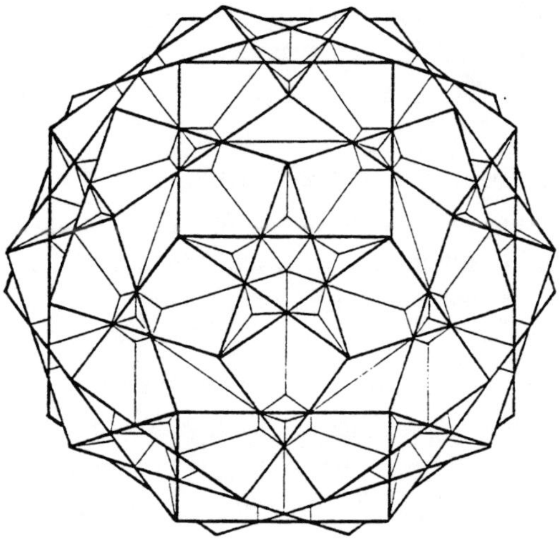

3 2 5/2| $t\begin{Bmatrix}3\\5/2\end{Bmatrix}$
[2.4g]
Truncated
Great Icosidodecahedron

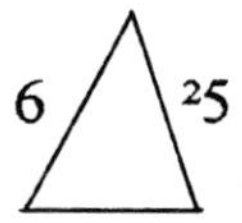

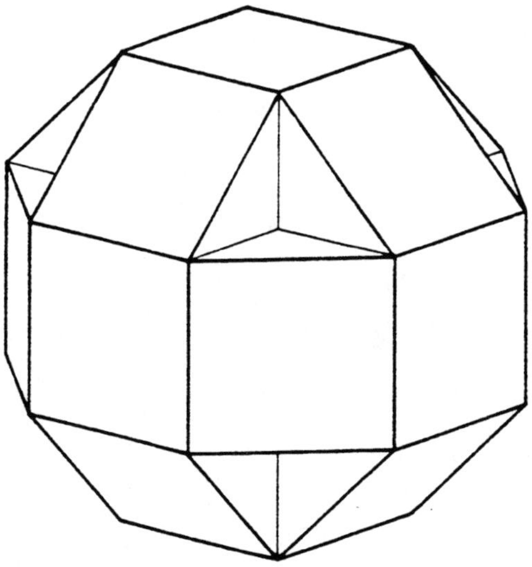

4 2 4/2| $t\begin{Bmatrix}4\\4/2\end{Bmatrix}$
[2.4m]
[1.3g]
Truncated Hexahexahedron

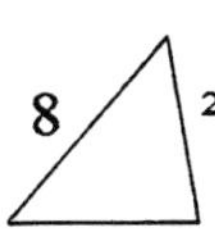

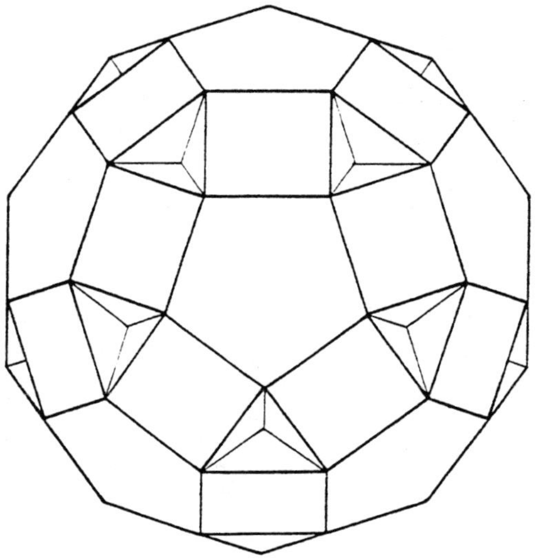

5 2 5/2| $t\begin{Bmatrix}5\\5/2\end{Bmatrix}$
[2.4j]
Truncated
Dodecadodecahedron

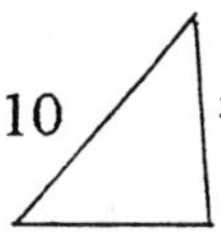

Fig.18: Even Faced Cross & Star Polyhedra

STAR RHOMBIHEDRA

We have already seen the polyhedra produced at all three vertices, the centre and two of the edges of our new Schwarz triangles.

That produced on the third edge, the 'hypotenuse', is as might be expected, a star rhombihedron, the second 'truncation' of the quasi-regular polyhedron in each family.

These semi-regular rhombihedra again have isosceles trapezoid vertex figures, the two parent polyhedra providing the opposed parallel sides, {x} and {n/2}, and two opposed squares {4} joining them together.

The star rhombihedron for each symmetry family is shown in Figure 19, and described numerically in Table 10 that follows.

3 5/2 |2 $r\begin{Bmatrix}3\\5/2\end{Bmatrix}$

Great
Rhombicosidodecahedron

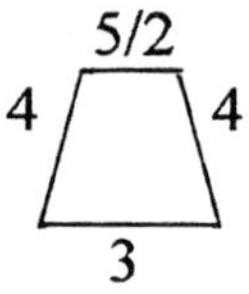

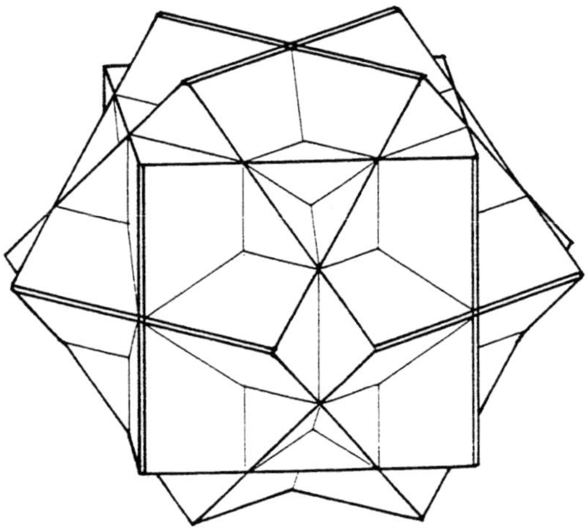

4 4/2 |2 $r\begin{Bmatrix}4\\4/2\end{Bmatrix}$

[1.3i]
Rhombihexahexahedron

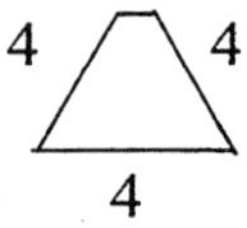

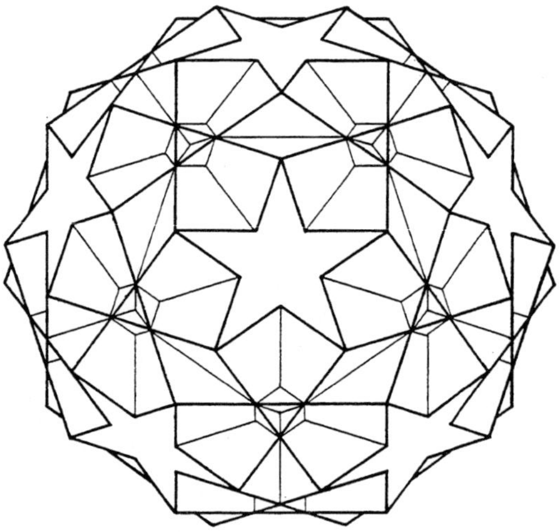

5 5/2 |2 $r\begin{Bmatrix}5\\5/2\end{Bmatrix}$

48 (76)
Rhombidodecadodecahedron

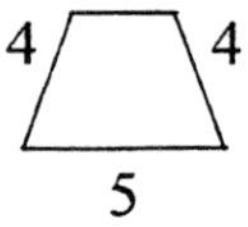

Fig.19: Cross & Star Rhombihedra

SNUB STAR POLYHEDRA

And so we come to the final polyhedron in our second series. As before, by taking the star rhombihedron, omitting the truncating square faces and twisting the remainder to leave room for two equilateral triangle faces in their place, we can produce a snub star polyhedron for each family.

These are shown below in Figure 20 and further described overleaf in Table 11. As with most snubs, they exist in two enantiomorphic mirror image forms that are differentiated by the direction of the twist.

That for the (4 2 4/2) family consists of three mutually perpendicular square antiprisms joined by the compound {4/2} faces. In fact, all of this family can be seen to be threefold compounds of the equivalent members of the (4 2 2) prism family.

|3 2 5/2 s{3 / 5/2}

73 (113)

Great Snub Icosidodecahedron

5/2 3 / 3 3 / 3

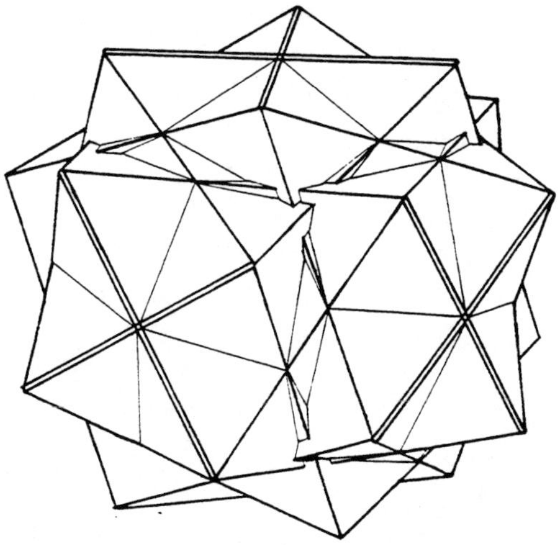

|4 2 4/2 s{4 / 4/2}

[1.3j]

Snub Hexahexahedron

4/2 3 / 3 3 / 4

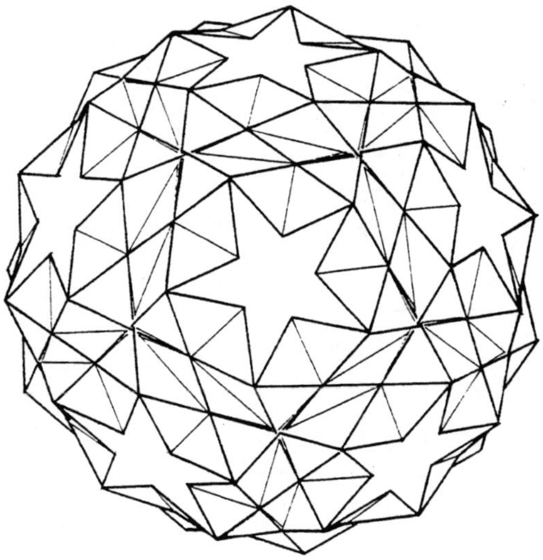

|5 2 5/2 s{5 / 5/2}

49 (111)

Snub Dodecadodecahedron

5/2 3 / 3 3 / 5

Fig.20: Snub Cross & Star Polyhedra

Symbol		V	E	F	
4 2 4/2|	t{ 4	24x2	24x2	6	⯃
[2.4m]	4/2}		24x1	6	⧈
Truncated Hexahexahedron				12	□
5 2 5/2|	t{ 5	60x2	60x2	12	⯃
[2.4j]	5/2}		60x1	12	⬠
Truncated Dodecadodecahedron				30	□
3 2 5/2|	t{ 3	60x2	60x2	20	⬡
[2.4g]	5/2}		60x1	12	⬠
Truncated Gt Icosidodecahedron				30	□

Table 9: Even Faced Cross & Star Polyhedra

4 4/2 |2	r{ 4	24	48	6	□
[1.3i]	4/2}			6	✕
Rhombihexahexahedron				12	□
5 5/2 |2	r{ 5	60	120	12	⬠
48 (76)	5/2}			12	☆
Rhombidodecadodecahedron				30	□
3 5/2 |2	r{ 3	20x3	60x2	20	△
	5/2}			12	☆
Great Rhombicosidodecahedron				30	□

Table 10: Cross & Star Rhombihedra

|4 2 4/2	s{ 4	24	60	6	□
[1.3j]	4/2}			6	✕
Snub Hexahexahedron				24	△
|5 2 5/2	s{ 5	60	150	12	⬠
49 (111)	5/2}			12	☆
Snub Dodecadodecahedron				60	△
|3 2 5/2	s{ 3	60	150	20	△
73 (113)	5/2}			12	☆
Great Snub Icosidodecahedron				60	△

Table 11: Snub Cross & Star Polyhedra

TRUNCATION & QUASITRUNCATION

The three symmetry families of cross and star polyhedra we have been examining have involved the polygons {5}, {4}, {3}, {5/2} and {4/2} along with their truncations {10}, {8}, {6}, $\{^{2}5/1\}$ and $\{^{2}4/1\}$.

We now wish to examine their performance under the action of quasitruncation.

The quasitruncations of these same polygons are truncations of {5/4}, {4/3}, {3/2}, {5/3} and {4/2} respectively, i.e. the conjugate polygons $\{^{2}5/2\}$, {8/3}, $\{^{2}3/1\}$, {10/3} and $\{^{2}4/1\}$, as shown below in Figure 21.

These have densities of 4, 3, 2, 3 and 2 respectively and comprise a double star, a star, a double polygon, a star and a double polygon in turn.

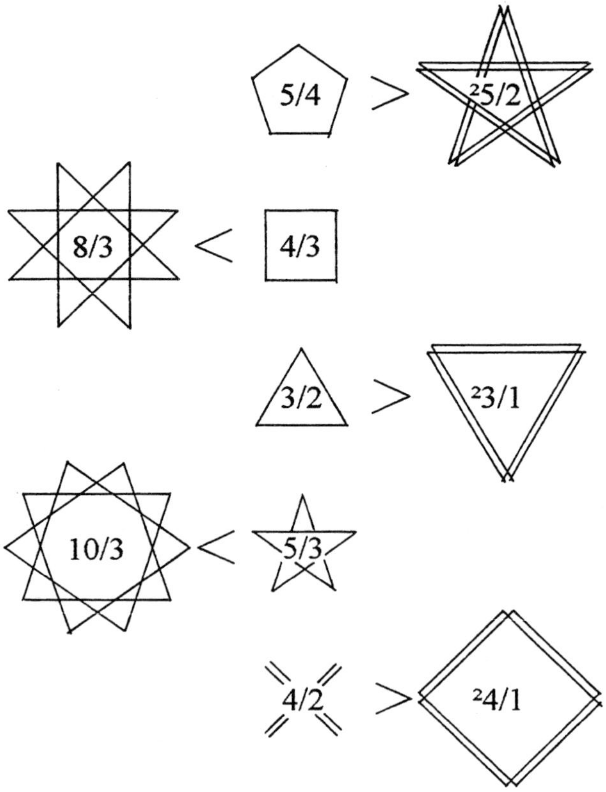

Fig.21: Quasitruncated Polygons

MULTIPLE VERTICES

Many of the polyhedra we have seen thus far have involved multiple faces, vertices or edges, often as a result of double polygon faces being introduced as the truncations of even denominator polygons. Most of these are summarised below in Figure 22, both as single vertices, and as multiple forms that give rise to the figures used in the tables.

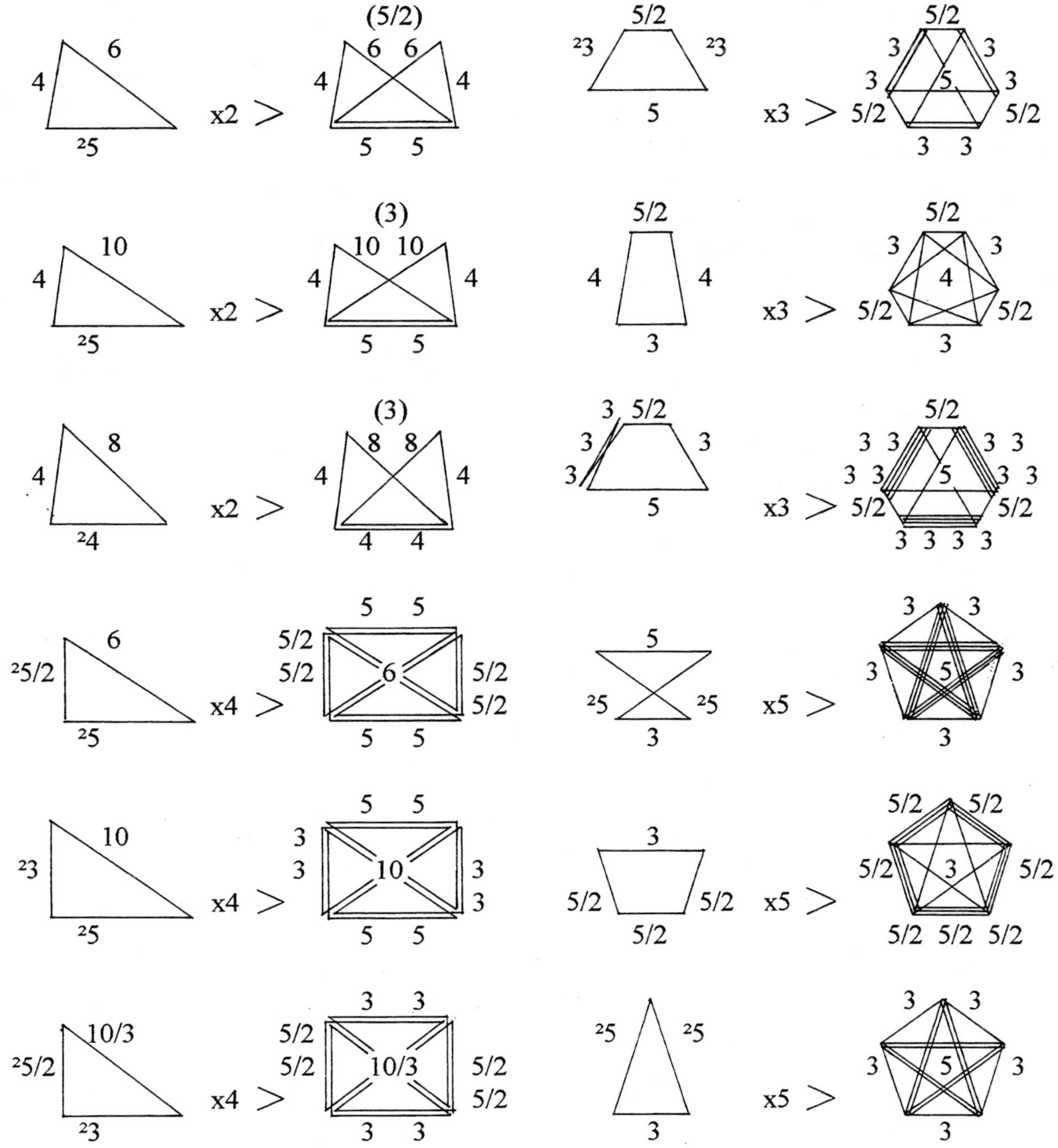

Fig.22: Multiple Vertex Figures

QUASITRUNCATED STAR POLYHEDRA

We saw in the previous volume, '*The Simpler? Polyhedra*' how a quasitruncated polyhedron can be obtained by substituting a quasitruncated polygon in place of a truncated one. This is in effect truncating a polyhedron made of inverse polygons.

The quasitruncated versions of our new regular cross and star polyhedra will thus have two quasitruncated faces and one truncating face at each vertex, again producing isosceles triangle vertex figures. These quasitruncated cross and star polyhedra are shown in Figures 23 & 24 and described below in Table 12.

Symbol	V	E	F
4/2 2 |4/3 t{4',4/2} [1.3c] Quasitruncated Great Hexahedron	24	36	6 6
4 2 |4/2 t{4/2',4} [1.3e] 3{4,3} Truncated Stellated Hexahedron	8x3	12x3	6 6
5/2 2 |5/4 t{5',5/2} 3{5/2,3} Quasitruncated Great Dodecahedron	20x3	30x3	12 12
5 2 |5/3 t{5/2',5} 74 (97) Quasitruncated Sm Stellated Dodecahedron	60	90	12 12
5/2 2 |3/2 t{3',5/2} {5/2,5} + 2{3,5/2} Quasitruncated Great Icosahedron	12x5	30x3	20 12
3 2 |5/3 t{5/2',3} 83 (104) Quasitruncated Gt Stellated Dodecahedron	60	90	12 20

Table 12: Quasitruncated Cross & Star Polyhedra

4/2 2 |4/3 t{4',4/2}
[1.3c]
Quasitruncated
Great Hexahedron

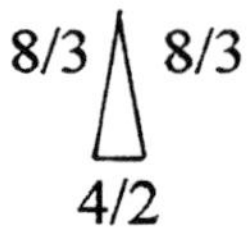

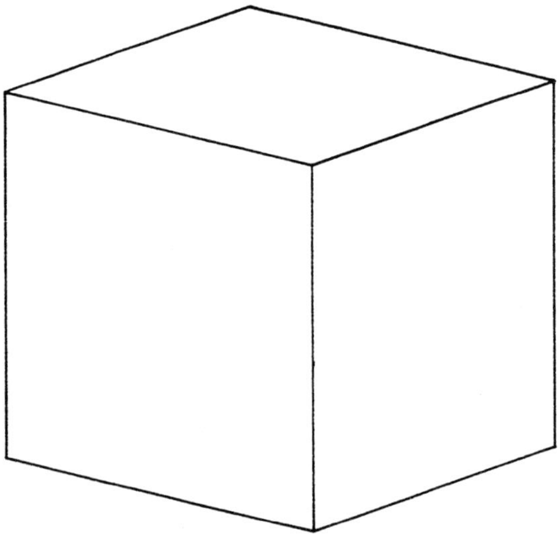

4 2 |4/2 t{4/2',4}
[1.3e] 3{4,3}
Truncated
Stellated Hexahedron

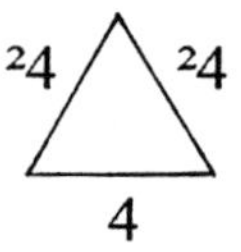

Fig.23: Quasitruncated Cross Polyhedra

5/2 2 |5/4 t{5',5/2}
3{5/2,3}
Quasitruncated
Great Dodecahedron

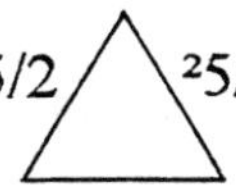

5/2 2 |3/2 t{3',5/2}
{5/2,5} + 2{3,5/2}
Quasitruncated
Great Icosahedron

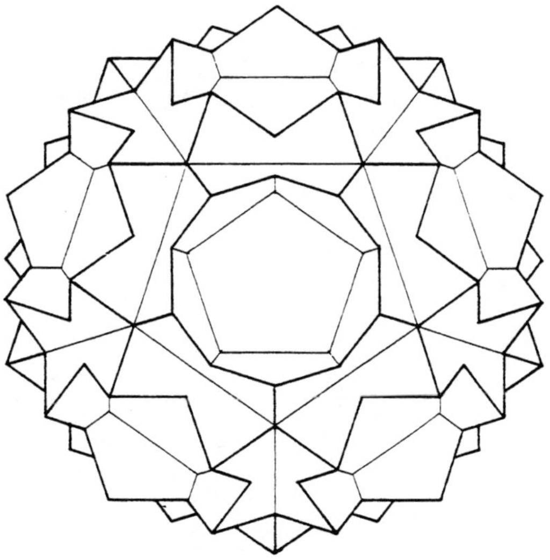

5 2 |5/3 t{5/2',5}
74 (97)
Quasitruncated
Small Stellated Dodecahedron

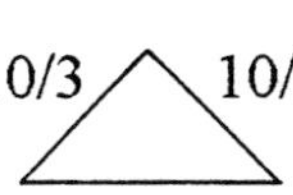

3 2 |5/3 t{5/2',3}
83 (104)
Quasitruncated
Great Stellated Dodecahedron

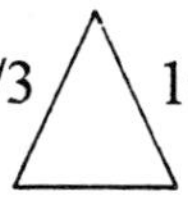

Fig. 24: Quasitruncated Star Polyhedra

QUASI EVEN FACED STAR POLYHEDRA

Continuing this quasitruncation process leads us back to our original quasi-regular polyhedron in each symmetry family. However these occur in two theoretical forms with different opposing pairs of inverse polygons around their rectangular vertex figures.

Truncating these in their turn and normalising the truncating faces back to squares as before, we can obtain two new quasitruncated even faced polyhedra for each family. These are shown in Figures 25 and 26 and described below in Table 13.

Symbol		V	E	F
4/3 2 4/2| [2.4n] Quasitruncated Hexahexahedron	t{4' / 4/2}	24x2	24x2 24x1	6 ✵ 6 ⧈ 12 □
4 2 4/2| [2.4m] Truncated Hexahexahedron	t{4/2' / 4}	24x2	24x2 24x2	6 ○ 6 ⧈ 12 □
5/4 2 5/2| [2.4k] Pentaquasitr. Dodecadodecahedron	t{5' / 5/2}	20x6	60x3	12 ★ 12 ⬠ 30 □
5 2 5/3| 75 (98) Quasitr. Dodecadodecahedron	t{5/2' / 5}	120	180	12 ○ 12 ✺ 30 □
3/2 2 5/2| [2.4h] Triquasitr. Gt Icosidodecahedron	t{3' / 5/2}	20x6	60x3	20 △ 12 ⬠ 30 □
3 2 5/3| 87 (108) Quasitr. Great Icosidodecahedron	t{5/2' / 3}	120	180	20 ⬡ 12 ✺ 30 □

Table 13: Quasi Even Faced Polyhedra

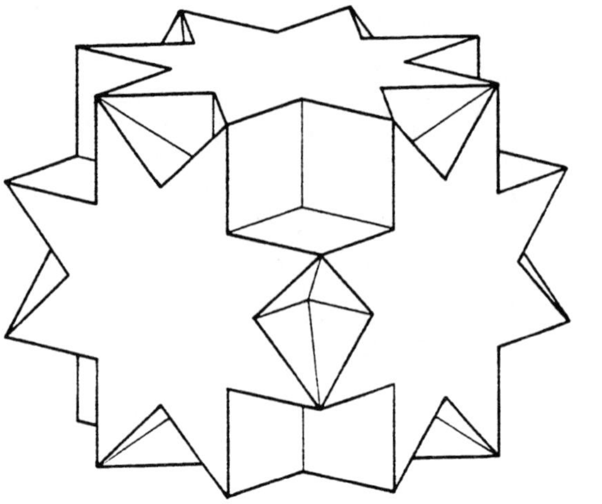

4/3 2 4/2| t{4' / 4/2}
[1.3h]
[2.4n]

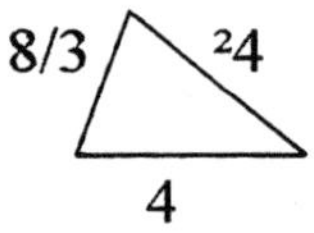

Quasitruncated Hexahexahedron

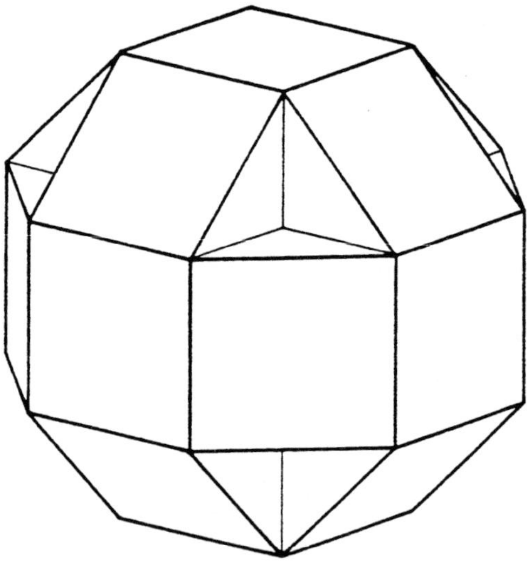

4 2 4/2| t{4/2' / 4}
[1.3g]
[2.4m]

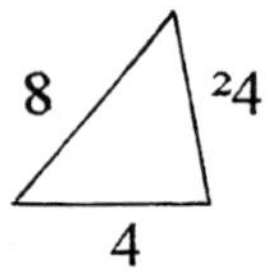

Truncated Hexahexahedron

Fig.25: Quasi Even Faced Cross Polyhedra

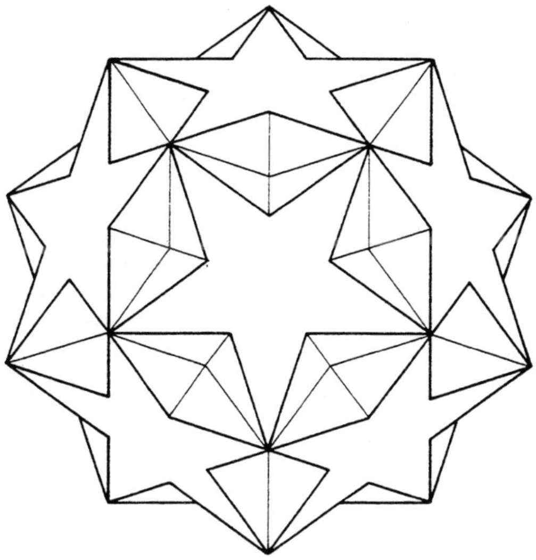

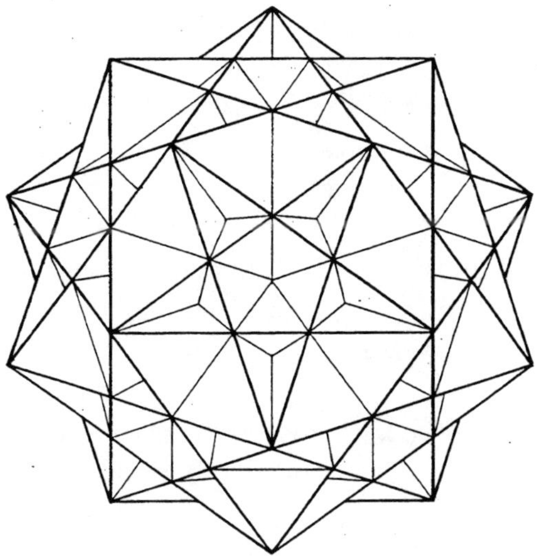

5/4 2 5/2| t{5' / 5/2}
[2.4k]
Pentaquasitruncated
Dodecadodecahedron

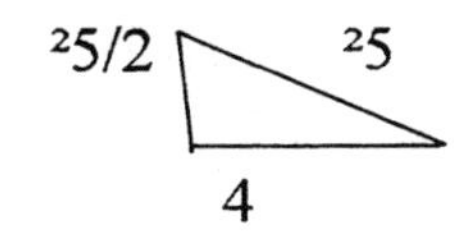

3/2 2 5/2| t{3' / 5/2}
[2.4h]
Triquasitruncated
Great Icosidodecahedron

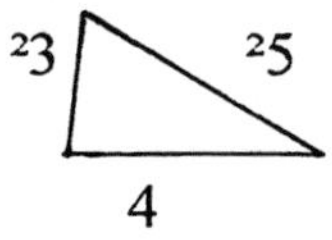

5 2 5/3| t{5/2' / 5}
75 (98)
Quasitruncated
Dodecadodecahedron

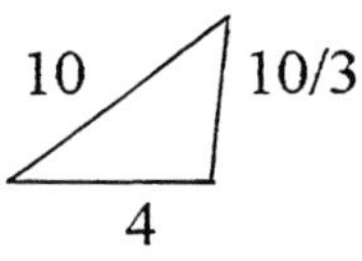

3 2 5/3| t{5/2' / 3}
87 (108)
Quasitruncated
Great Icosidodecahedron

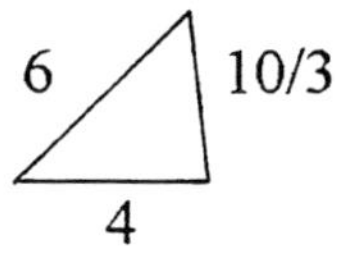

Fig.26: Quasi Even Faced Star Polyhedra

QUASIQUASITRUNCATED STAR POLYHEDRA

If we accept the existence of a fourth version of the quasi-regular polyhedron vertex figure, with both pairs of opposing faces being inverses, we can produce a doubly quasitruncated even faced polyhedron for each symmetry family.

These quasiquasitruncated cross and star polyhedra are the even faced polyhedra for the third colunar triangle, that off the 'hypotenuse' of the Schwarz triangle. They are shown below in Figure 27 and described in Table 14 that follows.

With both these and the quasitruncated versions on the previous pages, because {4/2} is self-inverse, its quasitruncation is the same as its truncation i.e. $\{^{2}4\}$, and the resultant polyhedron is one already seen.

3/2 2 5/3| t{ 3' / 5/2' }
[2.4i]
Quasiquasitruncated
Great Icosidodecahedron

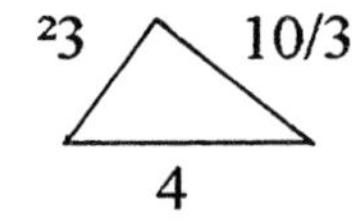

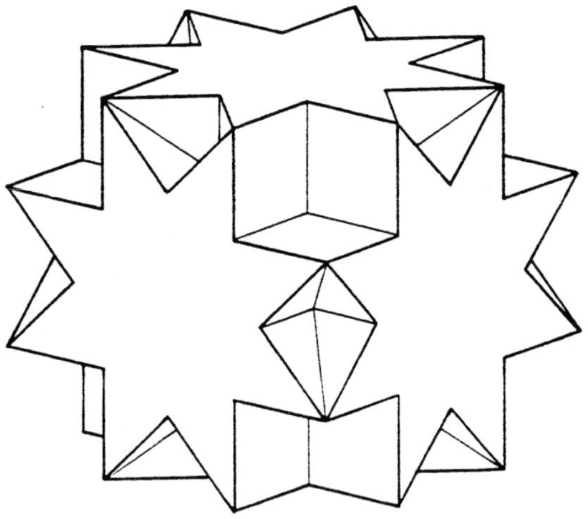

4/3 2 4/2| t{ 4' / 4/2' }
[2.4n]
[1.3h]
Quasitruncated Hexahexahedron

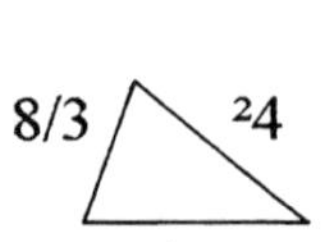

5/4 2 5/3| t{ 5' / 5/2' }
[2.4l]
Quasiquasitruncated
Dodecadodecahedron

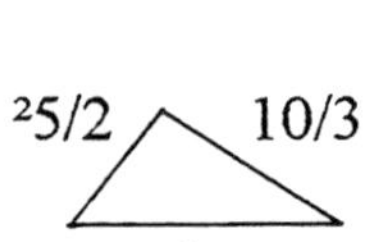

Fig.27: Quasiquasitruncated Cross & Star Polyhedra

STAR QUASIRHOMBIHEDRA

Continuing the quasitruncation process beyond the first 'truncation' quasitruncated even faced cross and star polyhedra we have just seen on pages 30 and 31, leads on to a second 'truncation' for each symmetry family. As with the regular polyhedra, these have two squares plus one normal and one inverse polygon opposing each other at each vertex.

Removing the inverse, by twisting the vertex figure to change its direction, causes the square faces to cross over each other. These cross and star quasirhombihedra are shown below in Figure 28 and described overleaf in Table 15.

Again the presence of the self-inverse cross polygon {4/2} leads to nothing new in the fourfold family.

3 5/3 |2 $r'\begin{Bmatrix}3\\5/2\end{Bmatrix}$ 5/2, 4 4, 3

84 (105)

Great Quasirhombicosidodecahedron

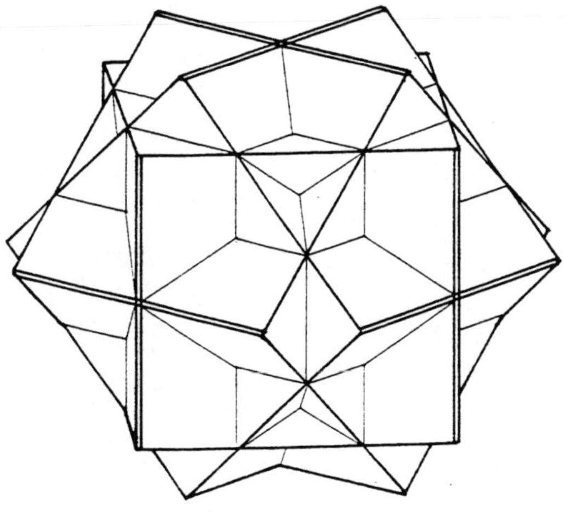

4 4/2 |2 $r'\begin{Bmatrix}4\\4/2\end{Bmatrix}$ 4/2, 4 4, 4

[1.3i]

Rhombihexahexahedron

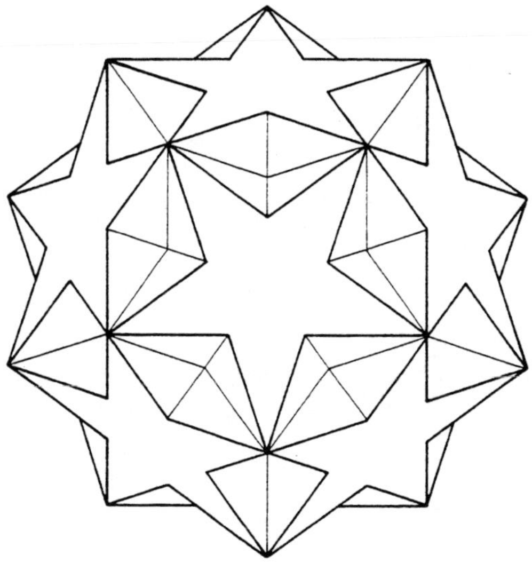

5 5/3 |2 $r'\begin{Bmatrix}5\\5/2\end{Bmatrix}$ 5/2, 4 4, 5

Quasirhombidodecadodecahedron

Fig.28: Cross & Star Quasirhombihedra

QUASISNUB STAR POLYHEDRA

Taking our star quasirhombihedra, we can omit the square faces and apply the usual twist to obtain quasisnub star polyhedra with paired triangular faces in lieu of the squares.

That for the (4 2 4/2) family is not shown here, as it is a repeat of the ordinary Snub Hexahexahedron already seen (p.25).

In its place, Figure 29 shows a second quasisnub polyhedron for the (3 2 5/2) family, obtained by treating the quasirhombihedron as 3/2 5/2 |2 rather than 3 5/3 |2.

Attempting this same trick for the (5 2 5/2) family should produce the quasisnub polyhedron |5/4 2 5/2, for which it is not possible to construct a vertex.

|3 2 5/3 s{3 / 5/2'}
88 (116)
Great Quasisnub
Icosidodecahedron

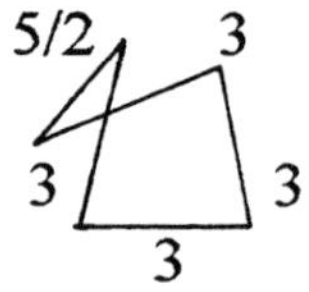

|3/2 2 5/2 s{3' / 5/2}
Small Quasisnub
Icosidodecahedron

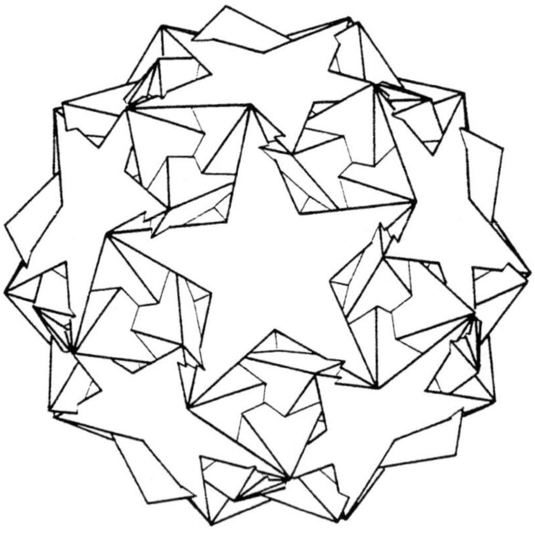

|5 2 5/3 s{5 / 5/2'}
76 (114)
Quasisnub
Dodecadodecahedron

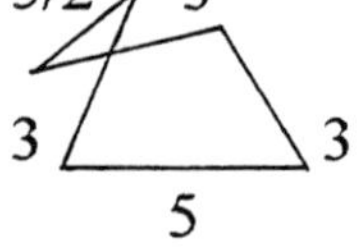

Fig.29: Quasisnub Star Polyhedra

Symbol		V	E	F	
4/3 2 4/2|	t{ 4' / 4/2' }	24x2	24x2	6	✡
[2.4n]			24x1	6	▣
Quasitruncated Hexahexahedron				12	□
5/4 2 5/3|	t{ 5' / 5/2' }	60x2	60x2	12	★
[2.4l]			60x1	12	✪
Qqtruncated Dodecadodecahedron				30	□
3/2 2 5/3|	t{ 3' / 5/2' }	60x2	60x2	20	△
[2.4i]			60x1	12	✪
Qqtruncated Gt Icosidodecahedron				30	□

Table 14: Quasiquasitruncated Polyhedra

4 4/2 |2	r'{ 4 / 4/2 }	24	48	6	□
[1.3i]				6	✕
Rhombihexahexahedron				12	□
5 5/3 |2	r'{ 5 / 5/2 }	20x3	60x2	12	⬠
				12	☆
Quasirhombidodecadodecahedron				30	□
3 5/3 |2	r'{ 3 / 5/2 }	60	120	20	△
84 (105)				12	☆
Gt Quasirhombicosidodecahedron				30	□

Table 15: Cross & Star Quasirhombihedra

|5 2 5/3	s{ 5 / 5/2' }	60	150	12	⬠
76 (114)				12	☆
Quasisnub Dodecadodecahedron				60	△
|3 2 5/3	s{ 3 / 5/2' }	60	150	20	△
88 (116)				12	☆
Gt Quasisnub Icosidodecahedron				60	△
|3/2 2 5/2	s{ 3' / 5/2 }	12x5	30x5	20	△
				12	☆
Sm Quasisnub Icosidodecahedron				60	△

Table 16: Quasisnub Star Polyhedra

P- & Q-TYPE STAR RHOMBIHEDRA

If we now take the star rhombihedra and quasirhombihedra for each of our new symmetry families, we can, as before, add and omit certain faces around the trapezoid vertex figures to obtain new polyhedra.

These are the p- and q-type rhombihedra together with the parallel p'- and q'-type quasirhombihedra.

Because the rhombihedron for the (4 2 4/2) family is the same as the quasirhombihedron and their p- and p'-types, only the q- (or q'-) type cross polyhedron is shown below in Figure 30.

The (5 2 5/2) and (3 2 5/2) star family equivalents are shown overleaf in Figures 31, 32, 33 and 34.

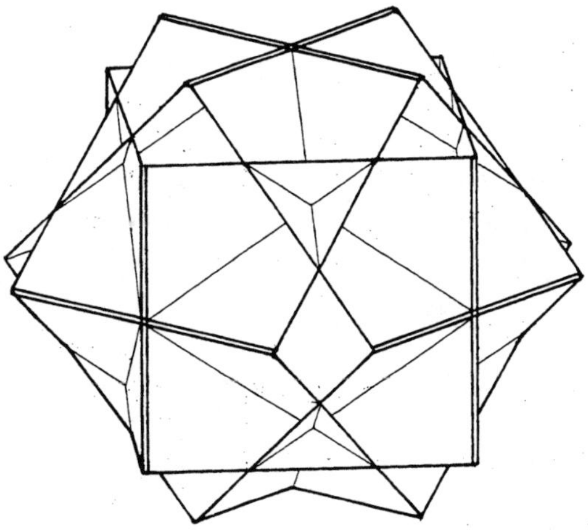

2 4/2 4/2| q{ 4 / 4/2 }
4/4|

Rhomboctahedron

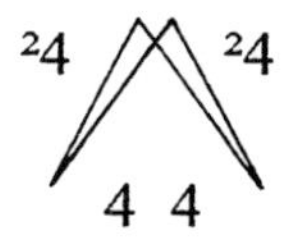

Fig.30: Q-type Cross Rhombihedron

5/3 3 |5/2 p{ 3 / 5/2 }

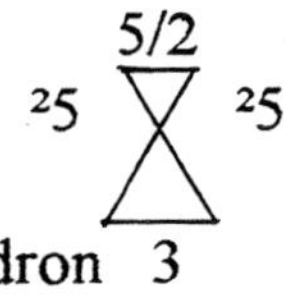

Great
Quasidodecicosidodecahedron

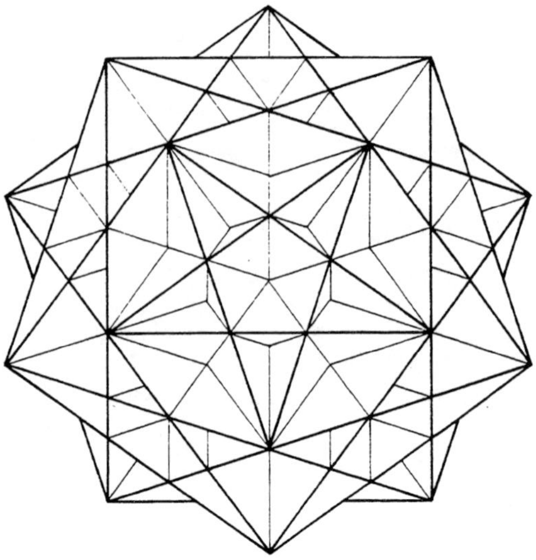

2 5/2 3/2|
5/4| q{ 3 / 5/2 }

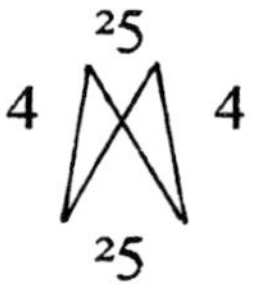

Great
Quasirhombidodecahedron

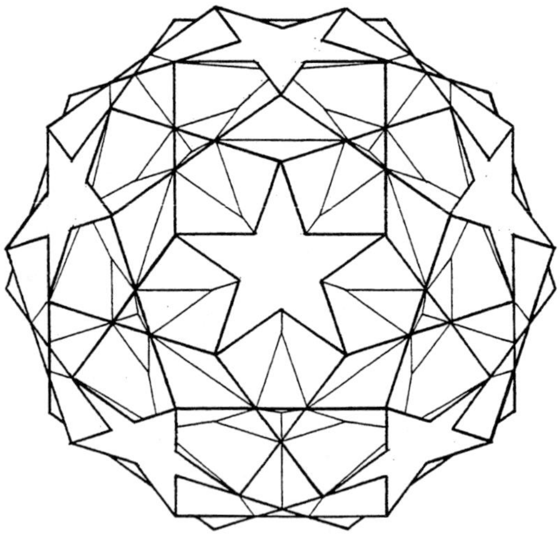

5/3 5 |3 p{ 5 / 5/2 }

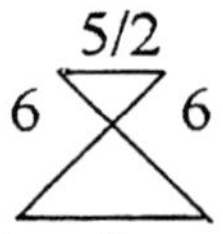

56 (83)
Icosidodecadodecahedron

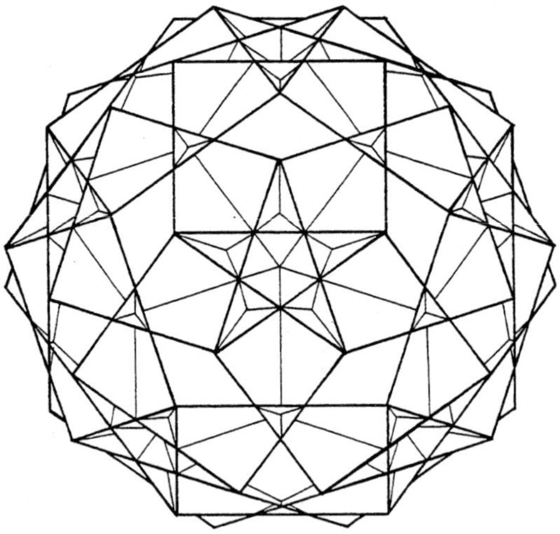

2 3 5/2|
5/4| q{ 5 / 5/2 }

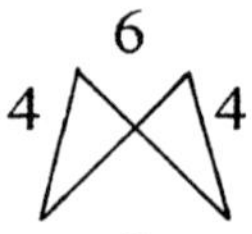

72 (96)
Rhombicosahedron

Fig.31: P-type Star Rhombihedra

Fig.32: Q-type Star Rhombihedra

3 5/2 |5/3 p'{3 / 5/2}
77 (99)
Great
Dodecicosidodecahedron

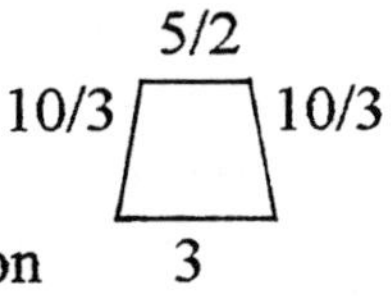

2 5/3 3/2| q'{3 / 5/2}
5/4|
89 (109)
Great Rhombidodecahedron

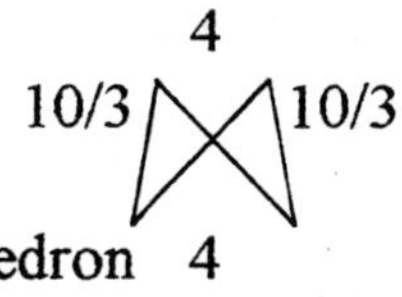

5 5/2 |3/2 p'{5 / 5/2}

Quasicosidodecadodecahedron

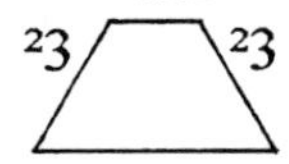

Fig.33: P'-type Star Quasirhombihedra

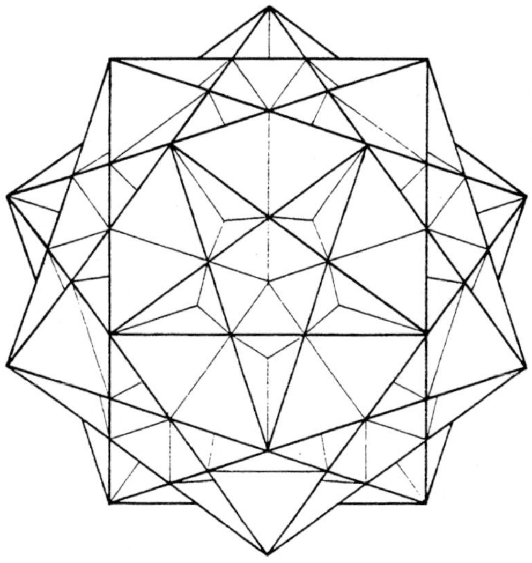

2 3/2 5/2| q'{5 / 5/2}
5/4|

Quasirhombicosahedron

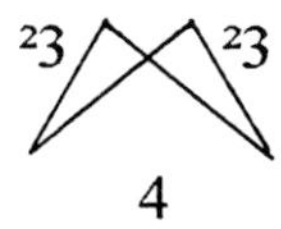

Fig.34: Q'-type Star Quasirhombihedra

THREE COMPLETE? STAR FAMILIES

In the preceding pages we have seen that our new cross and star polyhedra yield all the necessary members to be considered as three complete? new families of polyhedra: (4 2 4/2), (5 2 5/2) and (3 2 5/2).

These range from the regular and quasi-regular star polyhedra found at the vertices of the Schwarz triangles, through the truncated star polyhedra and star rhombi-hedra found on the edges, to the even faced star polyhedra found at the centre of each Schwarz triangle. In addition we have also seen the quasitruncated star polyhedra and star quasirhombihedra on the extended edges of the Schwarz triangles, together with the quasi even faced star polyhedra that occur at the centres of the colunar triangles.

These Schwarz triangles and their associated colunar triangles are shown for each family in the adjoining Figures 35, 36 and 37.

As we saw previously, the densities of these star polyhedra are 3, 3 or 7 accordingly, and this can be found to hold for quasitruncated versions as well, by simply treating the quasitruncated faces involved as negative.

An easy to understand example exhibiting this property is the Quasitruncated Small Stellated Dodecahedron of Figure 24, p. 29. This has between the outside and its centre one {10/3} quasitruncated face, one parallel {5} and five undercut {5} truncating faces, which gives a total density of (−3+1+5) = 3, as required for the (5 2 5/2) family.

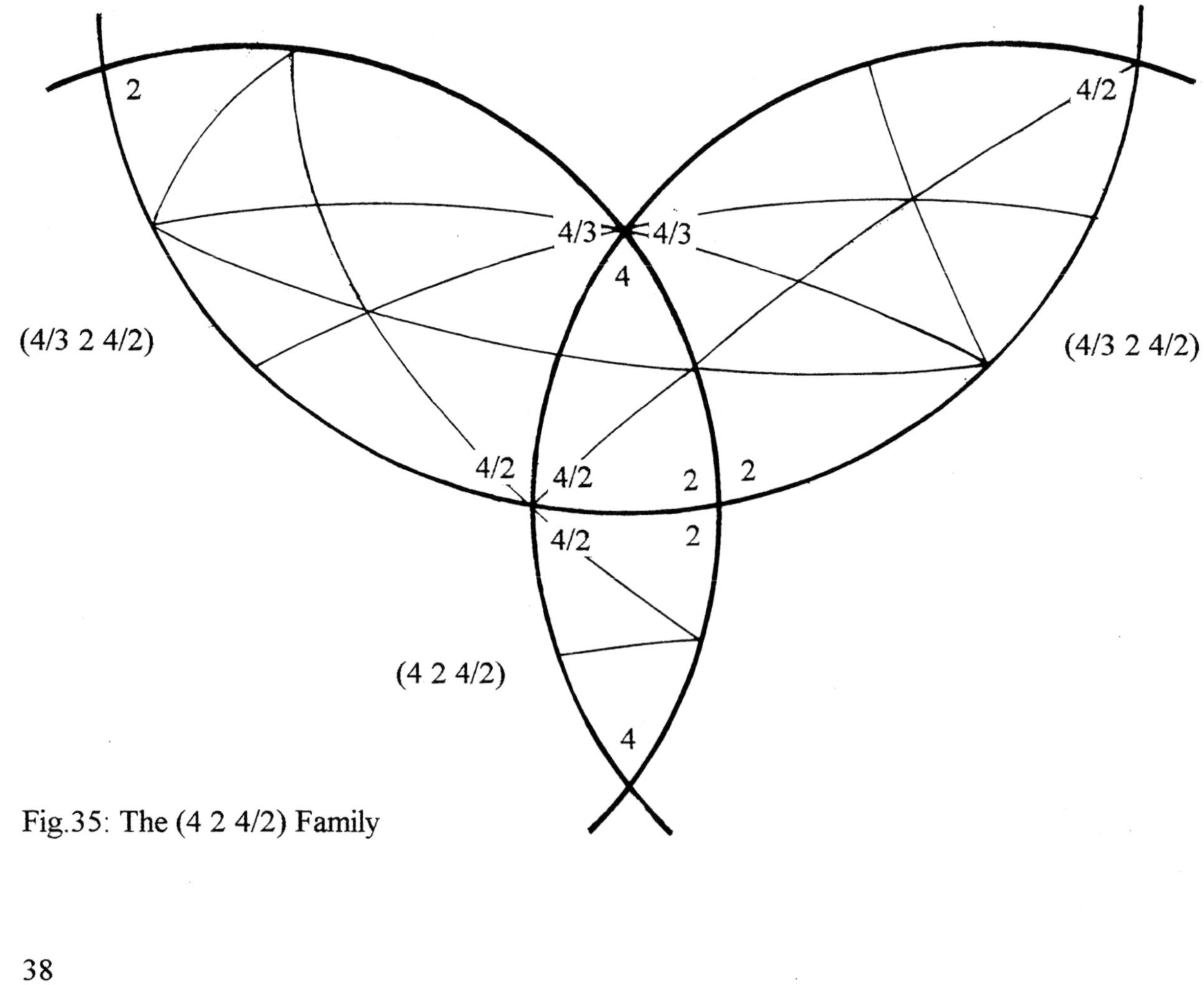

Fig.35: The (4 2 4/2) Family

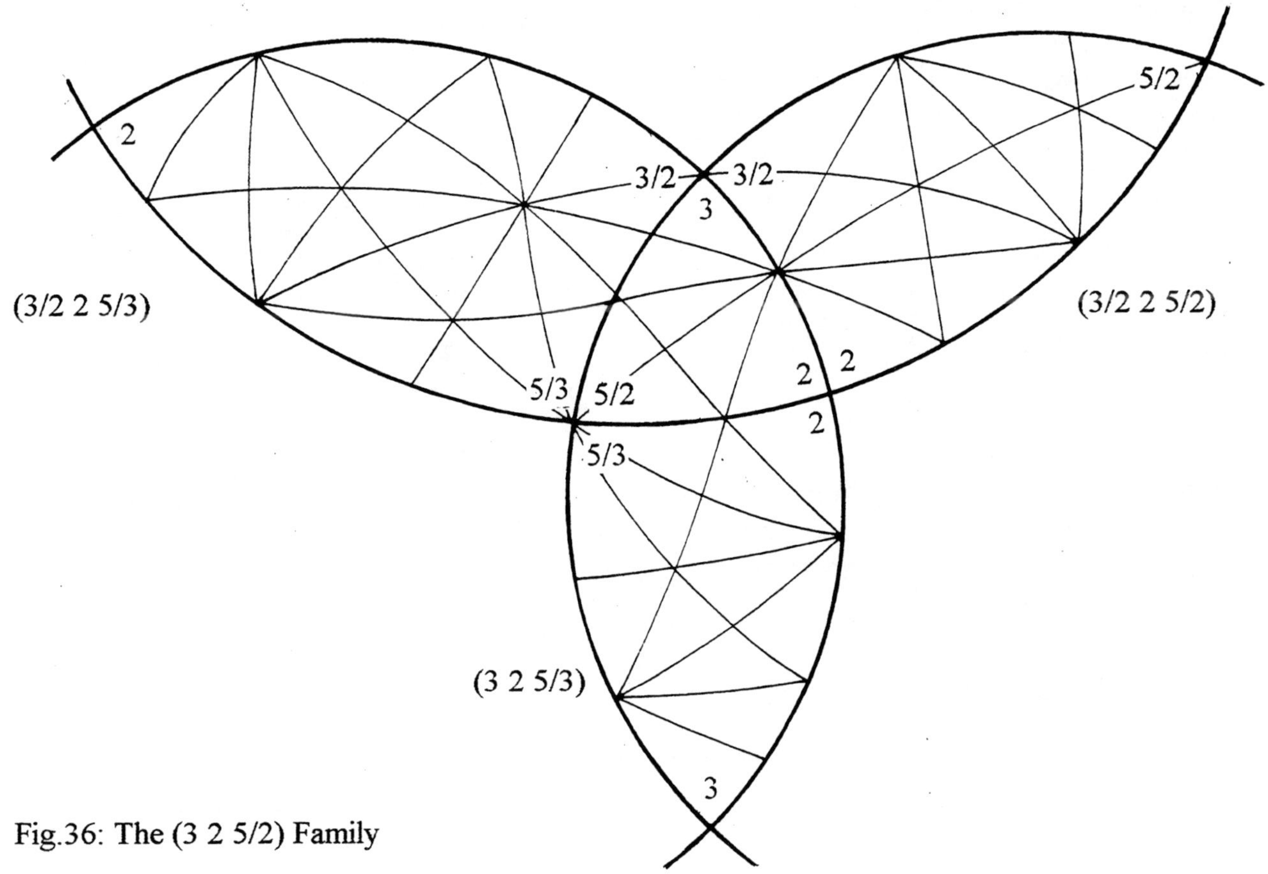

Fig.36: The (3 2 5/2) Family

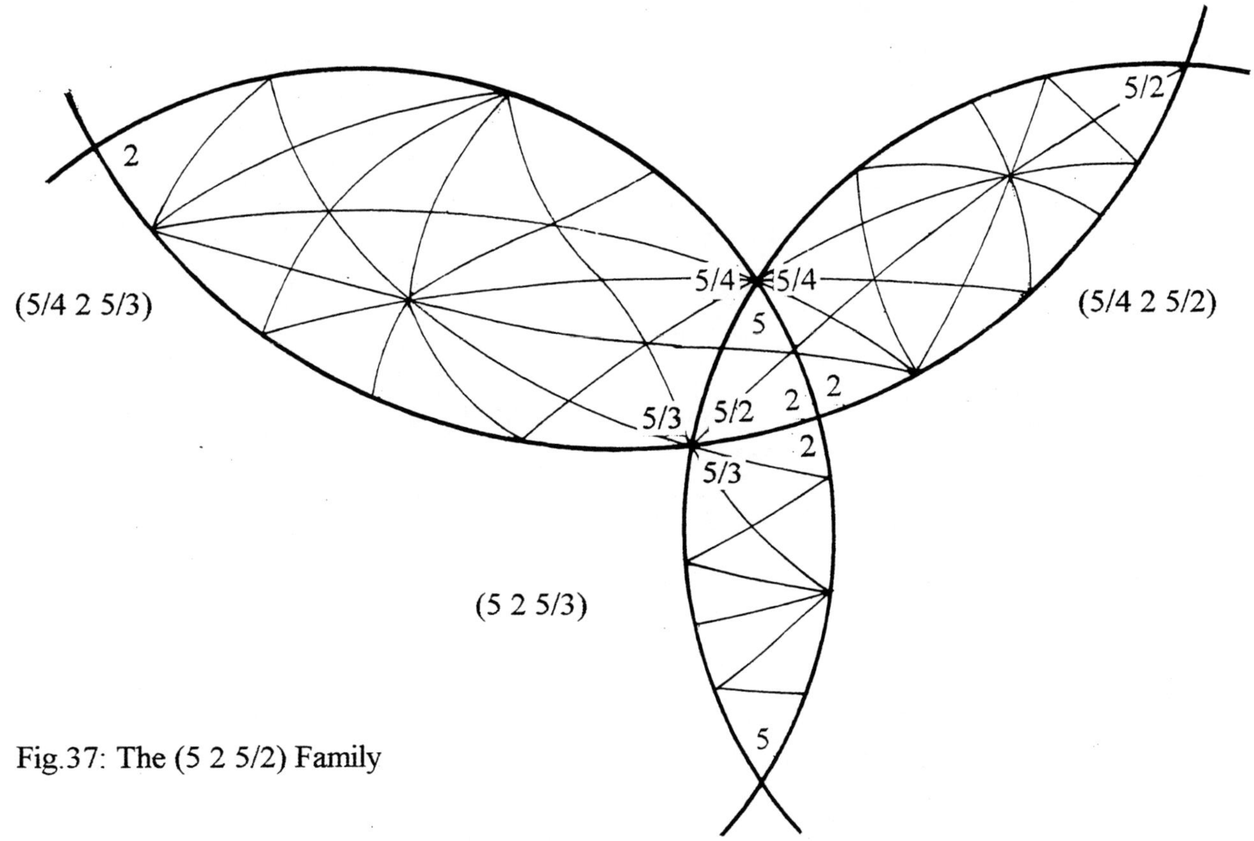

Fig.37: The (5 2 5/2) Family

FIVE MORE SCHWARZ TRIANGLES

By taking pairs of our three cross or star polyhedra Schwarz triangles and placing them back to back, we can form new Schwarz triangles for six theoretical sub-families.

These comprise (4 4/4 4), (4/2 4/2 4/2), (5 5/4 5), (5/2 5/2 5/2), (3 5/4 3) and (5/2 3/2 5/2).

The first of these is actually a complete lune, rather than a triangle, as it would include a theoretical {4/4} polygon at one vertex of the Schwarz triangle, giving an internal angle of 180° there.

The other five are shown in Figure 38 and form the basis of five sub-families related to our three original families of star polyhedra.

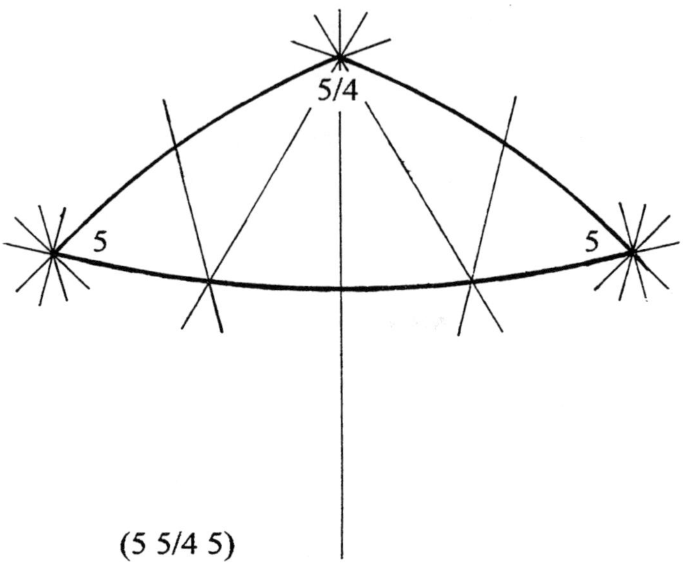

(5 5/4 5)

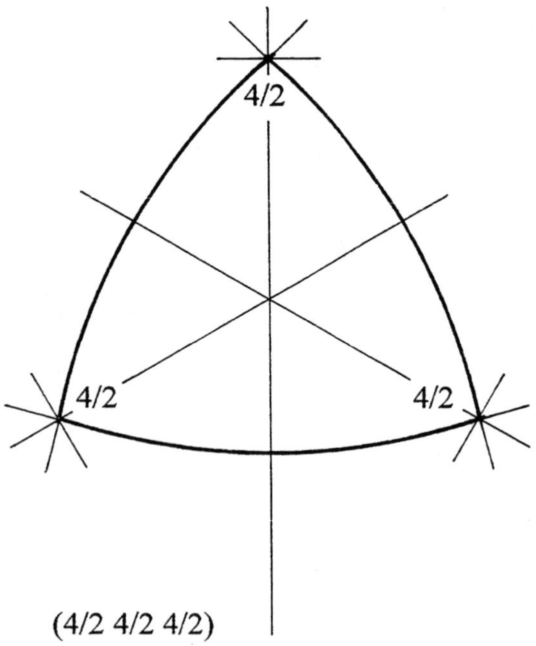

(4/2 4/2 4/2)

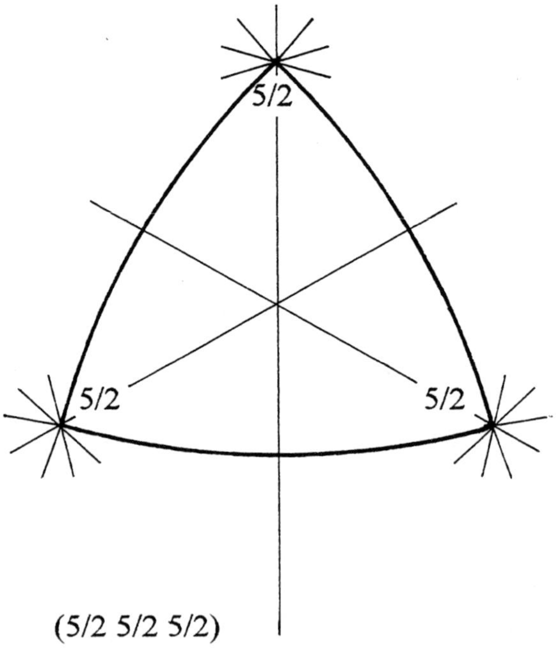

(5/2 5/2 5/2)

Fig.38: Sub-Family Schwarz Triangles

DOUBLE STAR POLYHEDRA

As with the previously seen sub-families, the polyhedra resulting from positions on the axis of symmetry of our new Schwarz triangles are simply double versions of those already seen.

Thus at the apex of the triangle we find a double version of a regular star polyhedron, in the centre we find an even faced star polyhedron which is a double version of a truncated star polyhedron, at the centre of the base we find a double version of a quasi-regular star polyhedron and finally in the centre of the lower colunar triangle we find another even faced star polyhedron, this time a double version of a quasitruncated star polyhedron.

These positions for typical star Schwarz triangles are shown below in Figure 39.

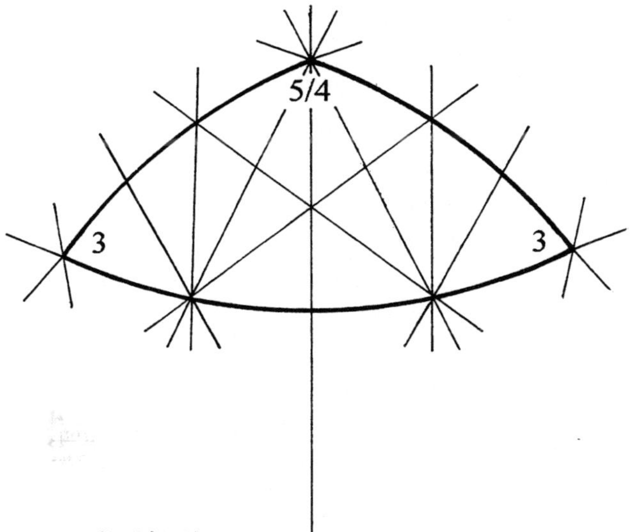

(3 5/4 3)

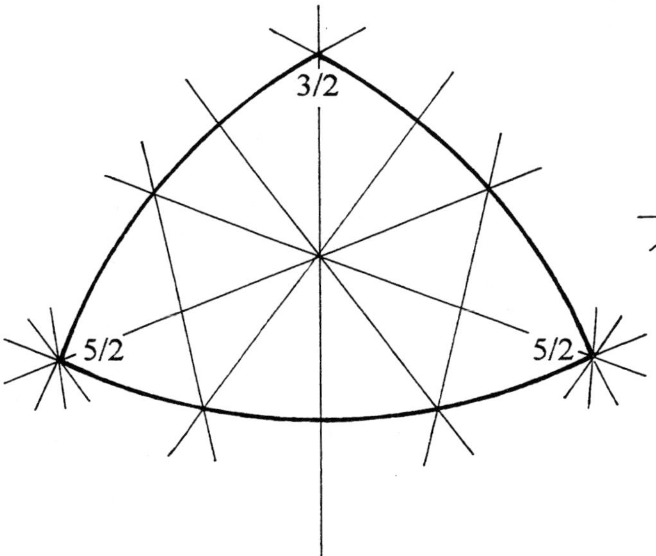

(5/2 3/2 5/2)

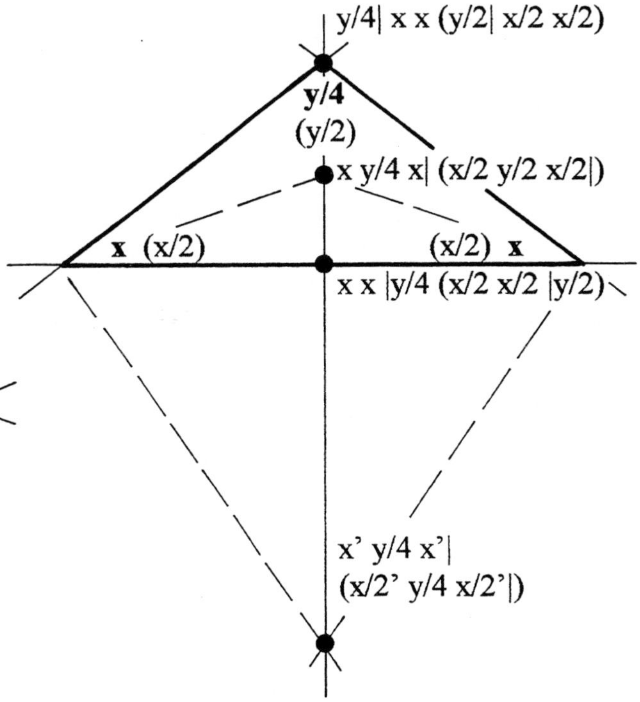

Fig.39: Double Star Polyhedra

MORE QUASI-REGULAR STAR POLYHEDRA

The double polyhedra we have just seen obviously have densities twice those of the star polyhedra of which they are double versions, i.e. 6 or 14 as appropriate.

The new Schwarz triangles will also yield further polyhedra with the same densities, as we shall see in the next few pages. If we start by taking either lower vertex of these sub-family triangles, we obtain a quasi-regular polyhedron for each sub-family.

The upper two of these shown below in Table 17 and Figure 40 adjoining, both involve a {5/4} generating polygon, which needs to be counted four times to make Euler's Formula work.

Similarly the lower three shown all involve three generating polygons of general type {x/2}, so that the vertices and faces they generate need to be counted twice.

Taking this into account we still get Euler's Formula to work in the general form V+F = E+2d, where d is the overall density of the polyhedron. This is 6 for the three polyhedra on the left, 14 for the pair on the right, matching the numbers of Möbius triangles in the respective Schwarz triangles.

For the former we thus have V+F=E+12, whilst for the latter V+F = E+28, which matches the results obtained for the double star polyhedra.

Symbol	V	E	F
3| 5/4 3	20	60	12 ⬠
61 (87)			20 △
Great Ditrigonal Icosidodecahedron			
5| 5/4 5	12	30x2	12 ⬠
2{5,5/2}			12 ⬠
Double Great Dodecahedron			
4/2| 4/2 4/2	6	12x2	6 ✕
2{4/2,4}			6 ✕
Double Stellated Hexahedron			
5/2| 5/2 5/2	12	30x2	12 ✯
2{5/2,5}			12 ✯
Double Small Stellated Dodecahedron			
5/2| 3/2 5/2	12	30x2	20 △
{5/2,5} + {3,5/2}			12 ✯
Inscribed Small Stellated Dodecahedron			

Table 17: Quasi-Regular Star Polyhedra

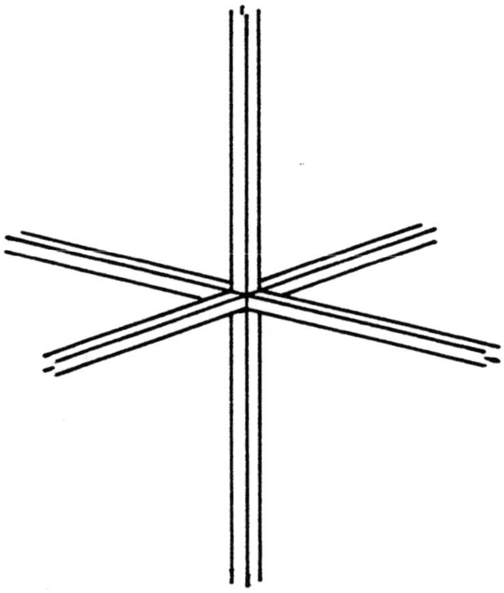

4/2| 4/2 4/2 — 4/2 □ 4/2 (4/2, 4/2)

2{4/2,4}

Double Stellated Hexahedron

Fig.40: More Quasi-Regular Star Polyhedra

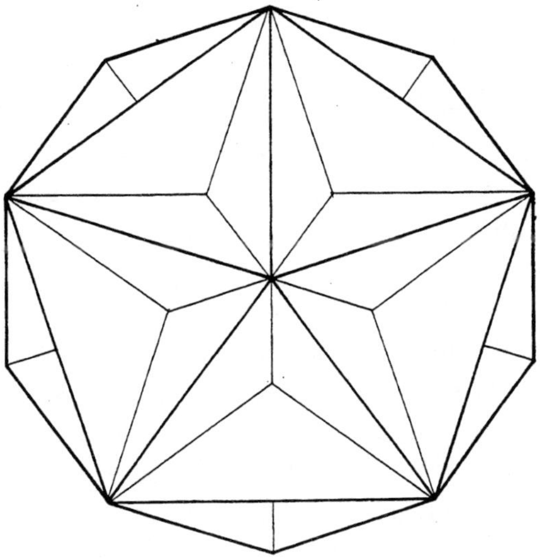

5| 5/4 5
2{5,5/2}
Double
Great Dodecahedron

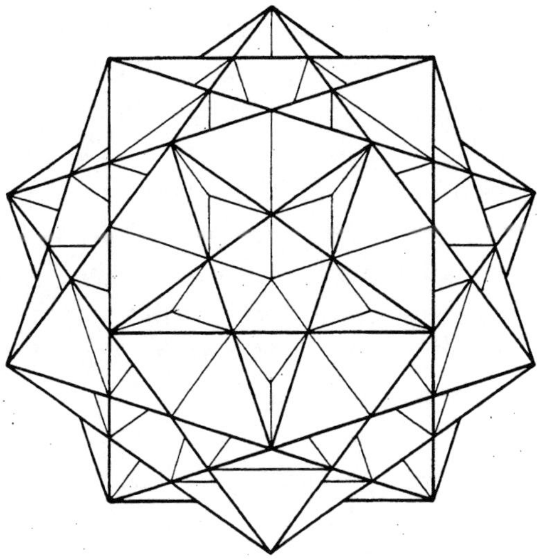

3| 5/4 3
61 (87)
Great
Ditrigonal Icosidodecahedron

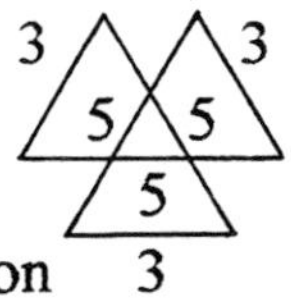

5/2| 5/2 5/2
2{5/2,5}
Double Small
Stellated Dodecahedron

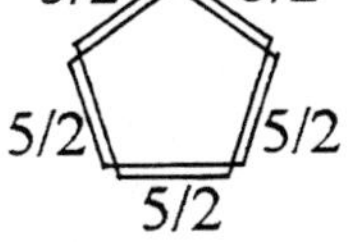

5/2| 3/2 5/2
{5/2,5} + {3,5/2}
Inscribed Small
Stellated Dodecahedron

MORE STAR RHOMBIHEDRA

Moving on around our new Schwarz triangles we will now examine the polyhedra generated on the upper edges. These are effectively star rhombihedra with trapezoid vertex figures, although three appear to be of quasirhombihedral form with twisted vertex figures caused by the inclusion of the inverse polygons {3/2} or {5/4}. In a similar way the equivalent star quasirhombihedra will in due course yield what appear to be rhombihedral forms (qv).

The two with straight rhombihedral type vertex figures, 4/2 4/2 |4/2 and 5/2 5/2 |5/2, are double versions of the quasi-regular star polyhedra. These are in fact those described on page 41, a result of the symmetry of their generating Schwarz triangles.

Of the other three, 5 5/4 |5, is actually a hemihedron and will be seen again later.

The remaining two on the right are both p-type rhombihedra, which have been seen before, one from the (3 5 5/3) super-family, the other from the (3 2 5/2) family of star polyhedra.

All of these polyhedra again have multiple density and can be made to comply with our revised Euler's Formulae, V+F=E+12 or V+F=E+28, by appropriate counting of the double density {x/2} faces twice, or the quadruple density {5/4} faces four times.

These star rhombihedra are shown in Figure 41 and described below in Table 18.

Symbol		V	E	F
3 5/4 |3 62 (88) Great Icosicosidodecahedron		60	120	12 ⬠ 20 △ 20 ⬡
5 5/4 |5 65 (91) Small Dodecahemidodecahedron	h{3 5'}	30x2	60x2	12 ⬠ 12 ⬠ 12 ○
4/2 4/2 |4/2 Double Hexahexahedron		12x2	24x2	6 ✕ 6 ✕ 6 ▣
5/2 5/2 |5/2 Double Dodecadodecahedron		30x2	60x2	12 ☆ 12 ☆ 12 ⬠
5/2 3/2 |5/2 Gt Quasidodecicosidodecahedron	p{3 5/2}	20x3	60x2	20 △ 12 ☆ 12 ⬠

Table 18: More Star Rhombihedra

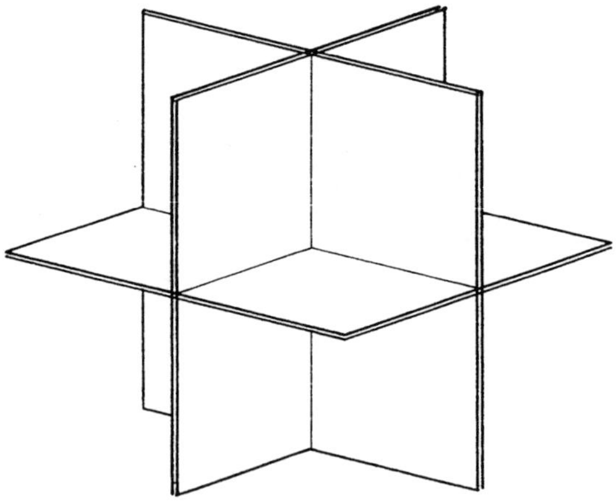

4/2 4/2 |4/2

Double Hexahexahedron

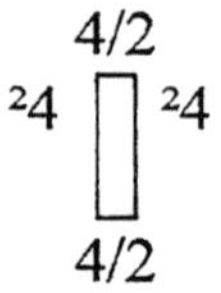

Fig.41: More Star Rhombihedra

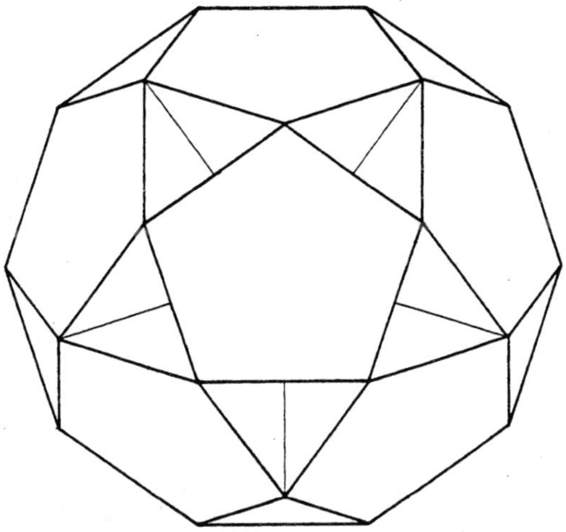

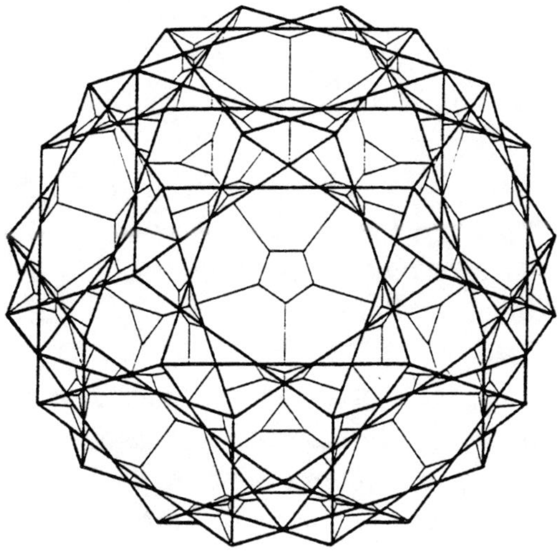

5 5/4 |5 $h\left\{\begin{matrix}3\\5'\end{matrix}\right\}$

65 (91)

Small

Dodecahemidodecahedron

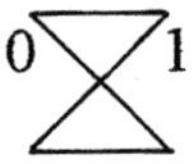

3 5/4 |3

62 (88)

Great Icosicosidodecahedron

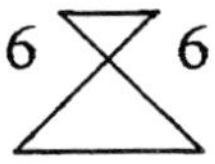

5/2 5/2 |5/2

Double Dodecadodecahedron

5/2 3/2 |5/2 $p\left\{\begin{matrix}3\\5/2\end{matrix}\right\}$

Great

Quasidodecicosidodecahedron

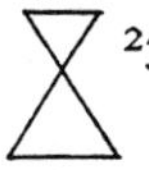

MORE EVEN FACED STAR POLYHEDRA

Having dealt with all positions within and around our new Schwarz triangles we will now take a look outside, more specifically at the centres of the upper colunar triangles.

Here we find further even faced polyhedra with the usual triangular vertex figures, as shown adjoining in Figure 42 and described below in Table 19.

Three of these are again double polygon versions of polyhedra we have already seen: two being truncations of the regular star polyhedra and one a double truncation of a regular cross polyhedron, itself a triple cube, hence we get a sixfold cube.

The other two also involve double polygons, which produce double triangle vertex figures, that fit within the shells of two previously seen rhombihedra, 3 5/2 |3 and 3 5 |5/3. These even faced polyhedra are in fact the complementary pair to the two even faced polyhedra derived in the (3 2 5) sub-families, as shown in Figure 37 on p.51 of '*The Simpler? Polyhedra*'.

All of these multiple density polyhedra still comply with Euler's formula, V+F=E+2d, and as all the faces in the vertex figures are truncations, including those that are double polygons, these polyhedra still count as even faced.

Symbol	V	E	F
3 5 3/2|	60x2	60x2	12
Quasiquasitruncated Great		60x1	20
Ditrigonal Icosidodecahedron			20
5 5 5/4|	60x2	90x2	12
Double Truncated			12
Great Dodecahedron			20
4/2 4/2 4/2|	8x6	12x6	6
Double Truncated			6
Stellated Hexahedron			6
5/2 5/3 5/3|	60x2	90x2	12
Double Quasitruncated			12
Small Stellated Dodecahedron			12
5/2 3 5/3|	60x2	60x2	20
Quasiquasitruncated		60x1	12
Inscr Sm Stellated Dodecahedron			12

Table 19: More Even Faced Star Polyhedra

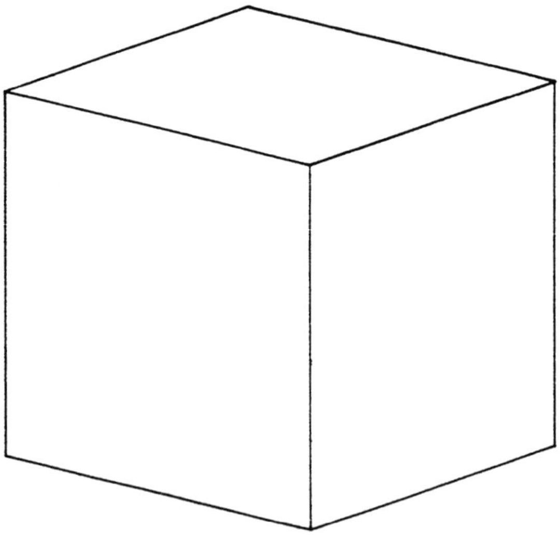

4/2 4/2 4/2|
6{4,3}
Double Truncated
Stellated Hexahedron

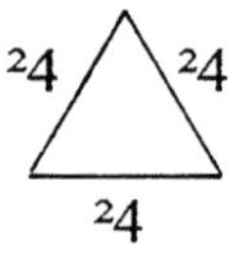

Fig.42: More Even Faced Star Polyhedra

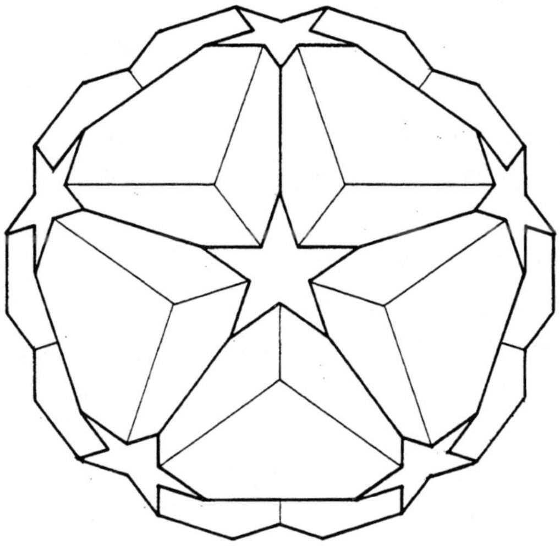

5 5 5/4|

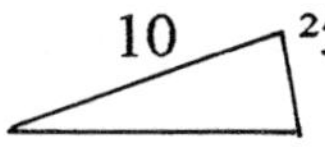

Double Truncated
Great Dodecahedron

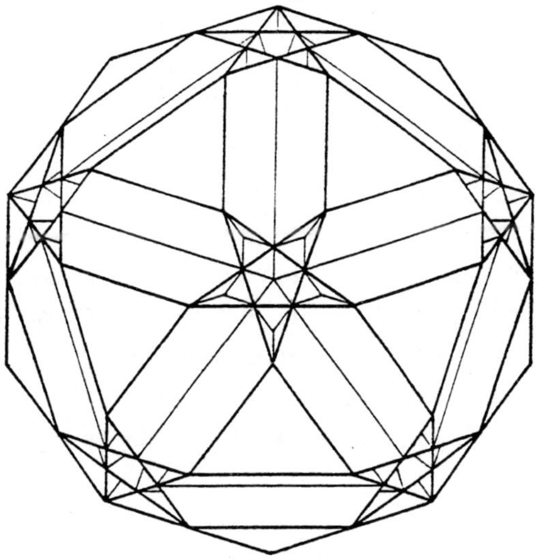

3 5 3/2|

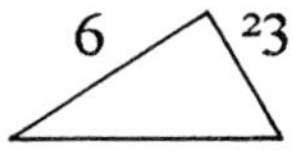

Quasiquasitruncated Great
Ditrigonal Icosidodecahedron

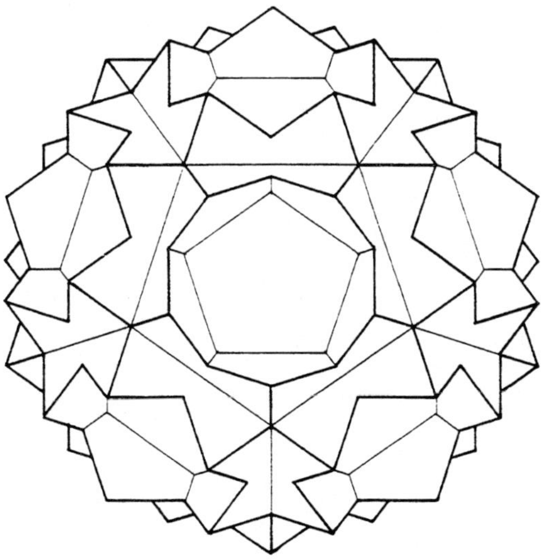

5/2 5/3 5/3|

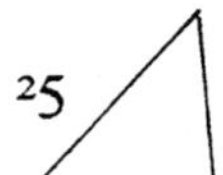

Double Quasitruncated
Small Stellated Dodecahedron

5/2 3 5/3|

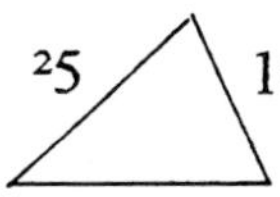

Quasiquasitruncated
Inscribed Sm Stellated Dodecahedron

MORE STAR QUASIRHOMBIHEDRA

Continuing our journey around the star sub-family Schwarz triangles we now come to examine the polyhedra produced on the extended upper edges of the triangles.

These are further star quasirhombihedra as shown in Figure 43 adjoining and described below in Table 20. Three do not however have the usual twisted vertex figures, since the polygon started with was an inverse. Taking its inverse in turn we end up with a regular polygon and a rhombihedral type vertex figure without a twist.

As with the equivalent star rhombihedra we again have on the left a hemihedron along with two double versions of quasi-regular star polyhedra.

Similarly the two on the right are p'-type quasirhombihedra, again from the (3 5 5/3) and (3 2 5/2) families, one obviously an inscribed version of the Icosahedron, whilst the Great Dodecicosidodecahedron below should perhaps include a 'quasi' in its name.

Because of the symmetry of the Schwarz triangles, the star polyhedra sub-families we have been examining have not so far yielded as many new polyhedra as did the regular polyhedra sub-families in the parallel section of the previous volume '*The Simpler? Polyhedra*'. This situation will be remedied somewhat in the next few pages where we examine the star hemihedra and the various snub polyhedra that can be derived in these sub-families.

Symbol	V	E	F
3 5 |3/2	12x5	30x4	12 ⬠
{5,5/2} + 3{3,5}			20 △
Great Quasicosicosidodecahedron			20 △
5 5 |5/4	30x2	60x2	12 ⬠
			12 ⬠
Double Dodecadodecahedron			12 ☆
4/2 4/2 |4/2	12x2	24x2	6 ✕
			6 ✕
Double Hexahexahedron			6 □
5/2 5/3 |5/3 h{3, 5/2'}	30x2	60x2	12 ☆
86 (107)			12 ☆
Great Dodecahemidodecahedron			12 ✡
5/2 3 |5/3 p'{3, 5/2}	60	120	20 △
77 (99)			12 ☆
Great Dodecicosidodecahedron			12 ✡

Table 20: More Star Quasirhombihedra

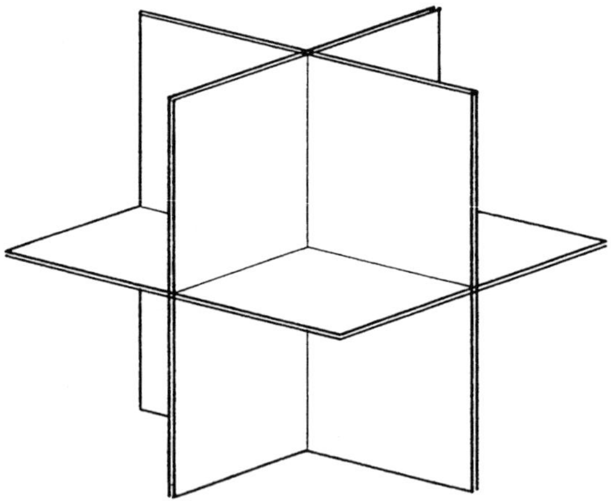

4/2 4/2 |4/2

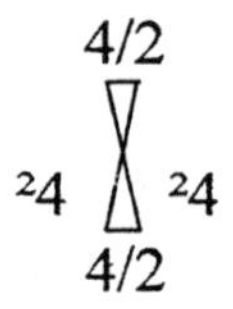

Double Hexahexahedron

Fig.43: More Star Quasirhombihedra

5 5 |5/4

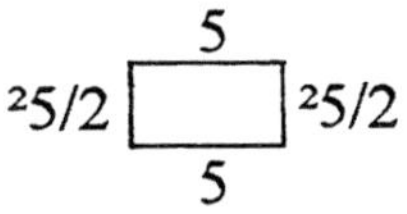

Double Dodecadodecahedron

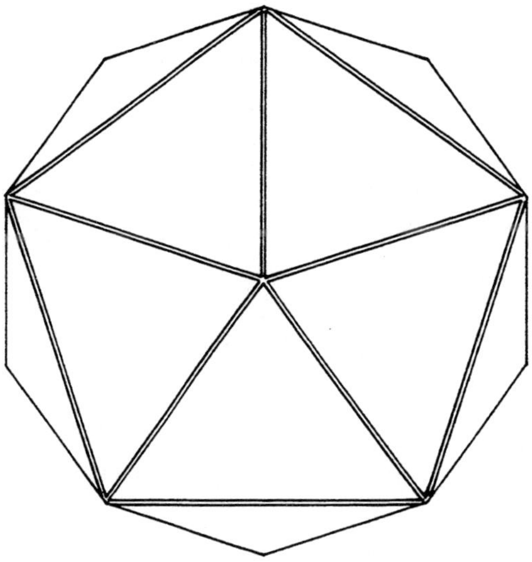

3 5 |3/2
{5,5/2} + 3{3,5}
Great
Quasicosicosidodecahedron

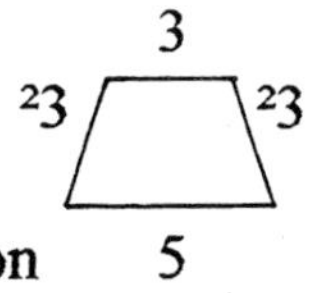

5/2 5/3 |5/3 h{ 3 }
86 (107) {5/2'}
Great
Dodecahemidodecahedron

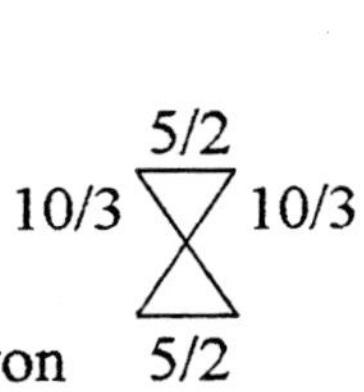

5/2 3 |5/3 p'{ 3 }
77 (99) {5/2}
Great
Dodecicosidodecahedron

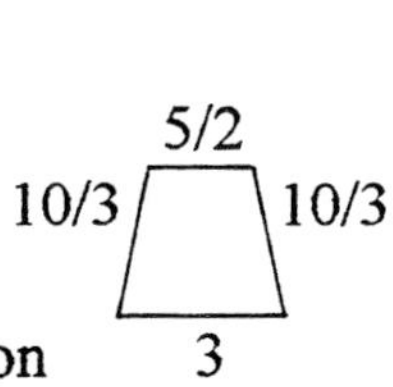

SOME STAR HEMIHEDRA

And so we come to the final position on our new sub-family Schwarz triangles, that on the extended lower edge. As with the regular polyhedra sub-families, this position yields hemihedra, polyhedra formed with an inverse pair of polygons opposing each other across the rectangular vertex figure, which consequently includes a twist.

The upper two seen in Figure 44, and described in Table 21 below, are the missing hemihedra from the (3 2 5) family, fitting within the shell of the Icosidodecahedron. The lower pair are the ‘other halves’ of those in the (3 2 5/2) and (5 2 5/2) families seen in the last volume. That for the (4 2 4/2) family is not illustrated, being yet another Double Hexahexahedron.

Symbol		V	E	F
3 3/2 |5 63 (89) Small Icosihemidodecahedron	h{3’ / 5}	30x2	60x2	12 ○ 20 △ 20 △
5 5/4 |5 65 (91) Small Dodecahemidodecahedron	h{3 / 5’}	30x2	60x2	12 ○ 12 ⬠ 12 ⬠
4/2 4/2 |4/2 Double Hexahexahedron	h{4/2’ / 4}	12x2	24x2	6 □ 6 ✕ 6 ✕
5/2 5/3 |5/3 86 (107) Great Dodecahemidodecahedron	h{3 / 5/2’}	30x2	60x2	12 ✡ 12 ☆ 12 ☆
5/2 5/3 |3 78 (100) Small Dodecahemicosahedron	h{5 / 5/2’}	30x2	60x2	20 ⬡ 12 ☆ 12 ☆

Table 21: Some Star Hemihedra

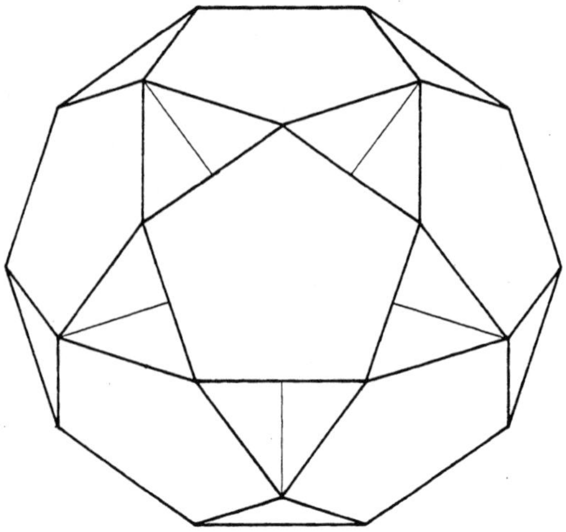

5 5/4 |5 h{3 / 5’}
65 (91)
Small
Dodecahemidodecahedron

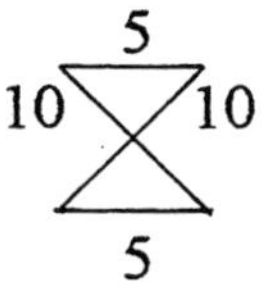

5/2 5/3 |5/3 h{3 / 5/2’}
86 (107)
Great
Dodecahemidodecahedron

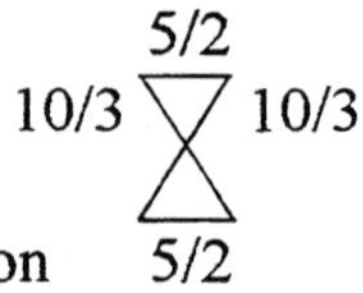

Fig.44: Some Star Hemihedra

MORE SNUB STAR POLYHEDRA

Taking the star rhombihedra on page 44 as the basis for snub polyhedra, we can form further vertex figures by taking the three sub-family polygons from their symbols and alternating them with triangles around a hexagonal vertex figure.

Note that for the (3 5/4 3) sub-family the snub cannot be constructed as a derivative of 3 5/4 |3, however the quasisnub star polyhedron derived from the same rhombihedron regarded as 3/2 5 |3 appears in the next section.

The remaining four are described below in Table 22 and shown overleaf in Figure 45. The two without inverse polygons are relatively simple, being threefold versions of the Stella Octangula, 3| 4/2 3, and Small Ditrigonal Icosidodecahedron, 3| 5/2 3.

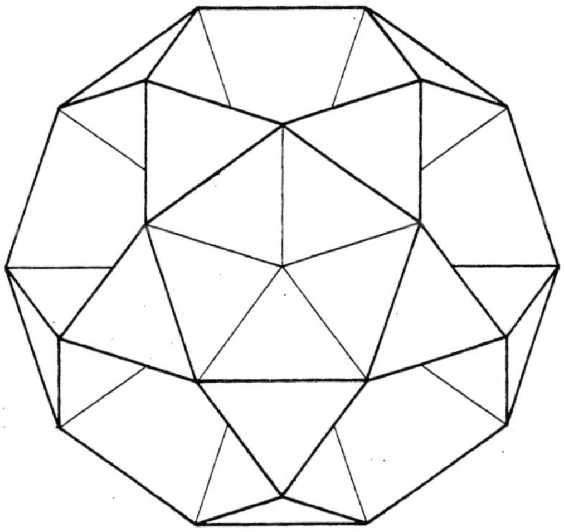

3 3/2 |5 h{3'}
63 (89) {5}
Small
Icosihemidodecahedron

5/2 5/3 |3 h{ 5 }
78 (100) {5/2'}
Small
Dodecahemicosahedron

Symbol	V	E	F
|3 5/4 3	not possible		
|5 5/4 5	12x5	30x6	12 ⬠
3{3,5} + 3{5,5/2}			24 ⬠
Triple Inscribed Icosahedron			60 △
|4/2 4/2 4/2	8x3	24x3	6 ✶
(19) [1.4a]			12 ✶
Triple Stella Octangula			24 △
|5/2 5/2 5/2	20x3	60x3	12 ☆
39 (70)			24 ☆
Triple Sm Ditr Icosidodecahedron			60 △
|5/2 3/2 5/2	30x2	60x3	20 △
70 (94)			24 ☆
Great Icosidodecahedron			60 △

Table 22: More Snub Star Polyhedra

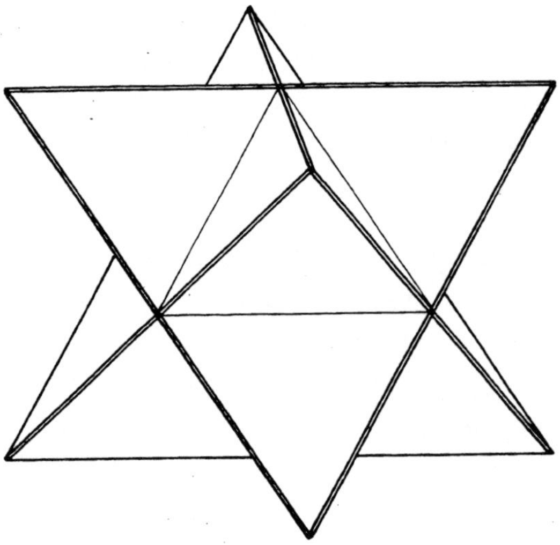

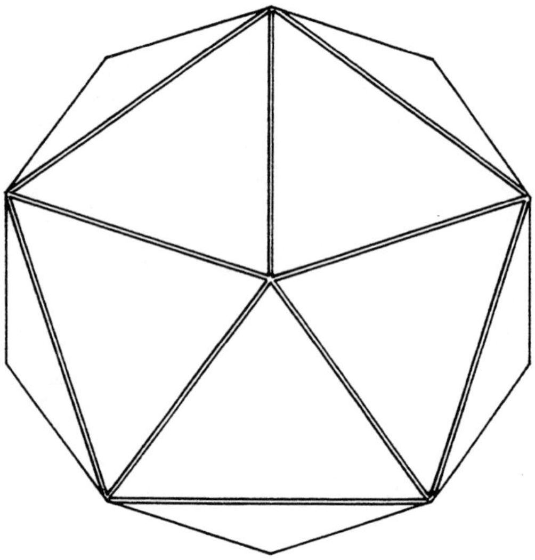

|4/2 4/2 4/2

(19) [1.4a]

Triple Stella Octangula

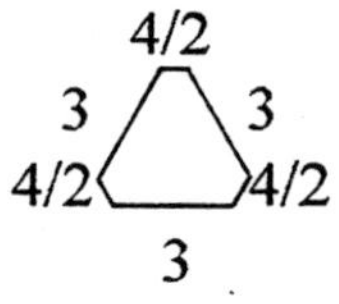

|5 5/4 5

3{3,5} + 3{5,5/2}

Triple Inscribed Icosahedron

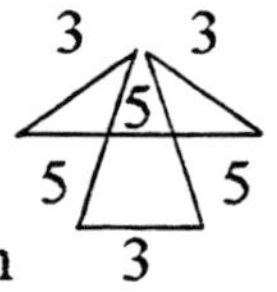

|5/2 5/2 5/2

39 (70)

Triple Small

Ditrigonal Icosidodecahedron

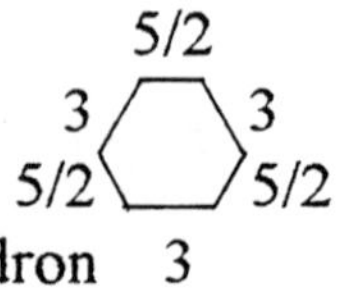

|5/2 3/2 5/2

70 (94)

Great Icosidodecahedron

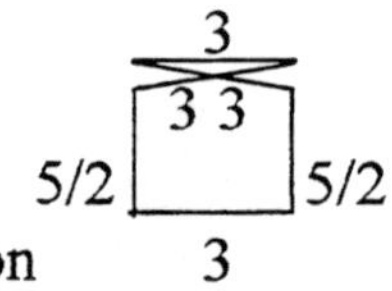

Fig.45: More Snub Star Polyhedra

MORE QUASISNUB STAR POLYHEDRA

If we perform a similar operation to that just carried out, but instead to the star quasi-rhombihedra from page 48, we arrive at the quasisnub star polyhedra.

These are described below in Table 23 and two of them are illustrated in Figure 46. The other three comprise |5 5 5/4 and |4/2 4/2 4/2, which we have already seen as ordinary snubs (Figure 45 top facing), along with |5/2 5/3 5/3, which will appear in the next section as a retrosnub.

The Great Snub Dodecicosidodecahedron illustrated below is particularly interesting as it fits within a polyhedral shell consisting of twenty interlocked octahedra that we have not seen before (qv).

Symbol	V	E	F
|3 5 3/2	12x5	30x6	12 ⬠
5{3,5} + {5,5/2}			40 △
Gt Qusnub Icosicosidodecahedron			60 △
|5 5 5/4	12x5	30x6	12 ⬠
3{3,5} + 3{5,5/2}			24 ⬠
Triple Inscribed Icosahedron			60 △
|4/2 4/2 4/2	8x3	24x3	6 ✕
(19) [1.4a]			12 ✕
Triple Stella Octangula			24 △
|5/2 5/3 5/3	12x5	30x6	12 ☆
3{5/2,5} + 3{3,5/2}			24 ☆
Triple Insc Sm Stell Dodecahedron			60 △
|5/2 3 5/3	60	180	20 △
80 (115)			24 ☆
Gt Snub Dodecicosidodecahedron			60 △

Table 23: More Quasisnub Star Polyhedra

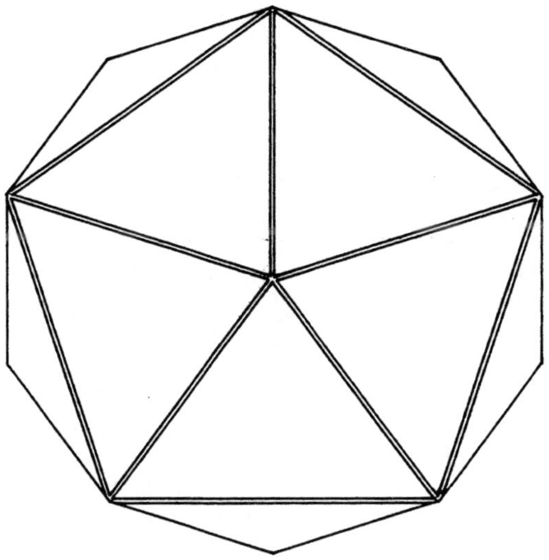

|3 5 3/2
5{3,5} + {5,5/2}
Great Quasisnub
Icosicosidodecahedron

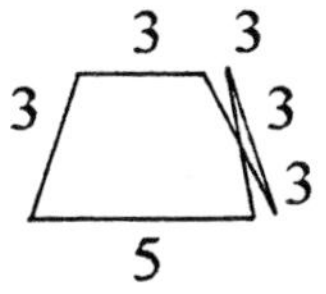

|5/2 3 5/3
80 (115)
Great Snub
Dodecicosidodecahedron

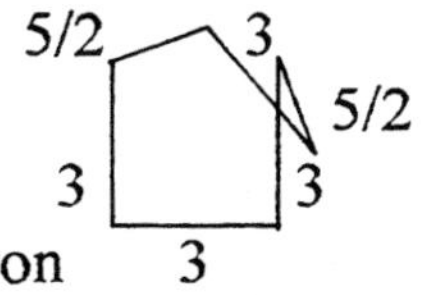

Fig.46: More Quasisnub Star Polyhedra

RETROSNUB STAR POLYHEDRA

Finally we come to the retrosnub star polyhedra, formed in the same way as the others but using the star quasirhombihedra in double inverse form as their generators.

These retrosnub polyhedra consequently have vertex figures with a double twist as shown in Figure 47 and described below in Table 24.

The five star sub-family members are here joined by a sixth retrosnub belonging to our original (3 2 5/2) star polyhedra family, the Great Retrosnub Icosidodecahedron.

Symbol	V	E	F	
|3/2 5/4 3/2	12x5	30x6	12	⬠
5{3,5} + {5,5/2}			40	△
Gt Retrosn Icosicosidodecahedron			60	△
|5/4 5/4 5/4	20x3	60x3	12	⬠
61 (87)			24	⬠
Triple Gt Dit Icosidodecahedron			60	△
|3/2 2 5/3	60	180	20	△
90 (117)			12	☆
Gt Retrosnub Icosidodecahedron			60	△
|4/2 4/2 4/2	8x3	24x3	6	✕
(19) [1.4a]			12	✕
Triple Stella Octangula			24	△
|5/3 5/2 5/3	12x5	30x6	12	☆
3{5/2,5} + 3{3,5/2}			24	☆
Tr Insc Sm Stell Dodecahedron			60	△
|5/3 3/2 5/3	12x5	30x6	20	△
2{5/2,5} + 4{3,5/2}			24	☆
Gt Retro Dodecicosidodecahedron			60	△

Table 24: Retrosnub Star Polyhedra

|3/2 2 5/3
90 (117)
Great Retrosnub
Icosidodecahedron

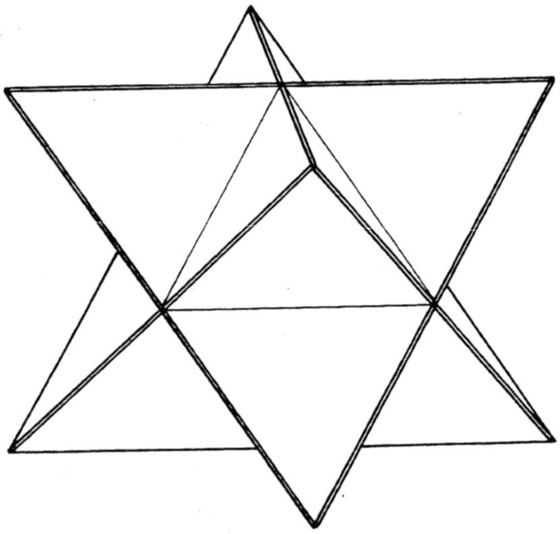

|4/2 4/2 4/2

(19) [1.4a]
Triple Stella Octangula

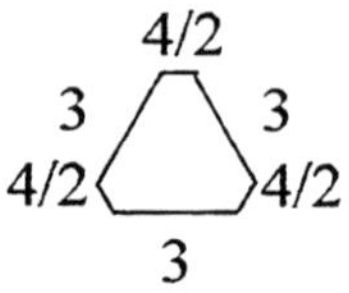

Fig.47: Retrosnub Star Polyhedra

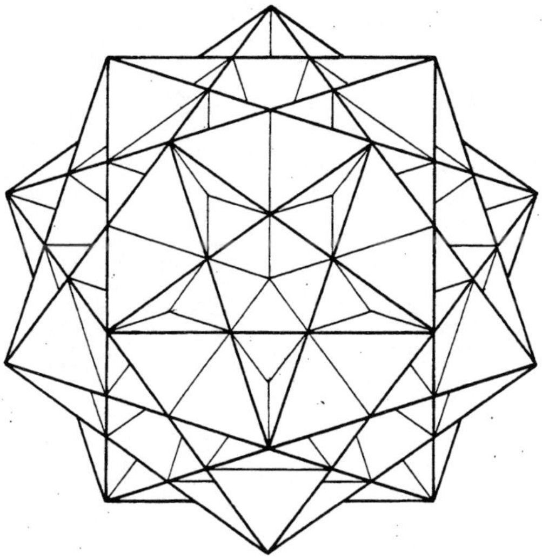

|5/4 5/4 5/4
61 (87)
Triple Great
Ditrigonal Icosidodecahedron

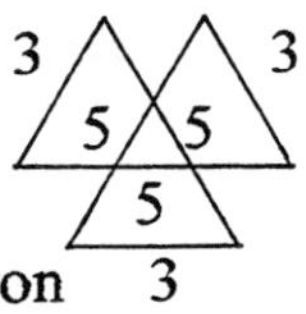

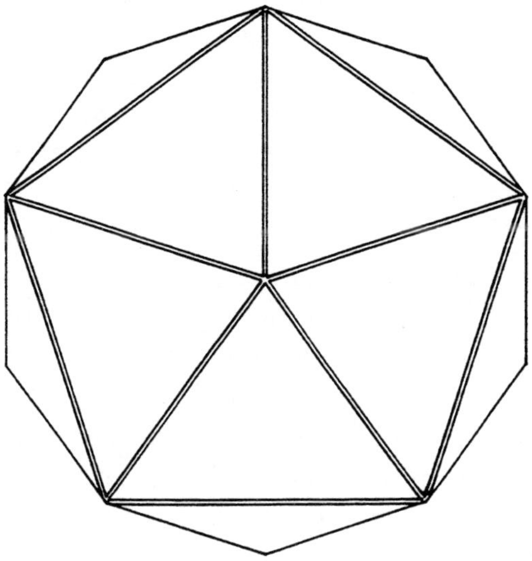

|3/2 5/4 3/2
5{3,5} + {5,5/2}
Great Retrosnub
Icosicosidodecahedron

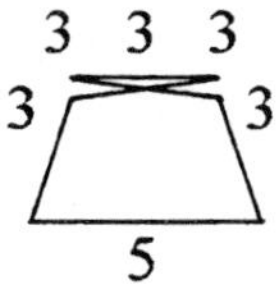

|5/3 5/2 5/3
3{5/2,5} + 3{3,5/2}
Triple Inscribed Small
Stellated Dodecahedron

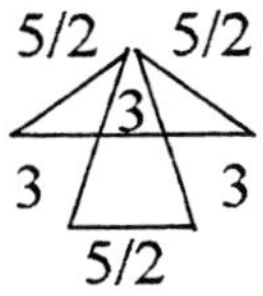

|5/3 3/2 5/3
2{5/2,5} + 4{3,5/2}
Great Retrosnub
Dodecicosidodecahedron

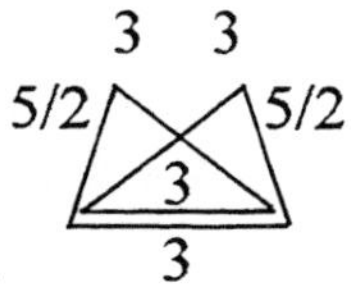

MORE MULTIPLE VERTICES

Since looking at multiple vertices back on page 27 we have encountered a great many more vertices with double polygons and multiple forms. These are shown adjoining in Figure 48, both in their single forms and as multiples.

In the latter, the number of times the single vertex repeats itself tells us how many times each vertex is visited in making up the complete polyhedron. Similarly each pair of edges meeting at a vertex of the vertex figure represents one edge between two faces of the polyhedron, telling us how many times each edge is visited to make the complete polyhedron. We thus obtain the multiples needed for inclusion in our tables which ensure that Euler's Formula continues to work.

Having thus adjusted our values of V and E, we then only have to make the appropriate adjustments to F, taking into account the densities of the faces. This gives us a general version of Euler's Formula in the revised form V+F=E+2d, where d is the overall density of the polyhedron, equivalent to the number of Möbius triangles in the polyhedron's respective family Schwarz triangle.

This can be seen to work for even the most complex retrosnub polyhedron, e.g. from Table 24 the Inscribed Icosahedron labelled as a Great Retrosnub Icosicosidodecahedron (|3/2 5/4 3/2) has twelve 5-fold vertices (60), twelve internal pentagon faces derived from {5/4} which are therefore counted four times (48) and one hundred triangular faces comprising the five coincident Icosahedra (100), a total V+F=208. It also has thirty 6-fold edges (180) and an overall family density of fourteen so that E+2d=208 too.

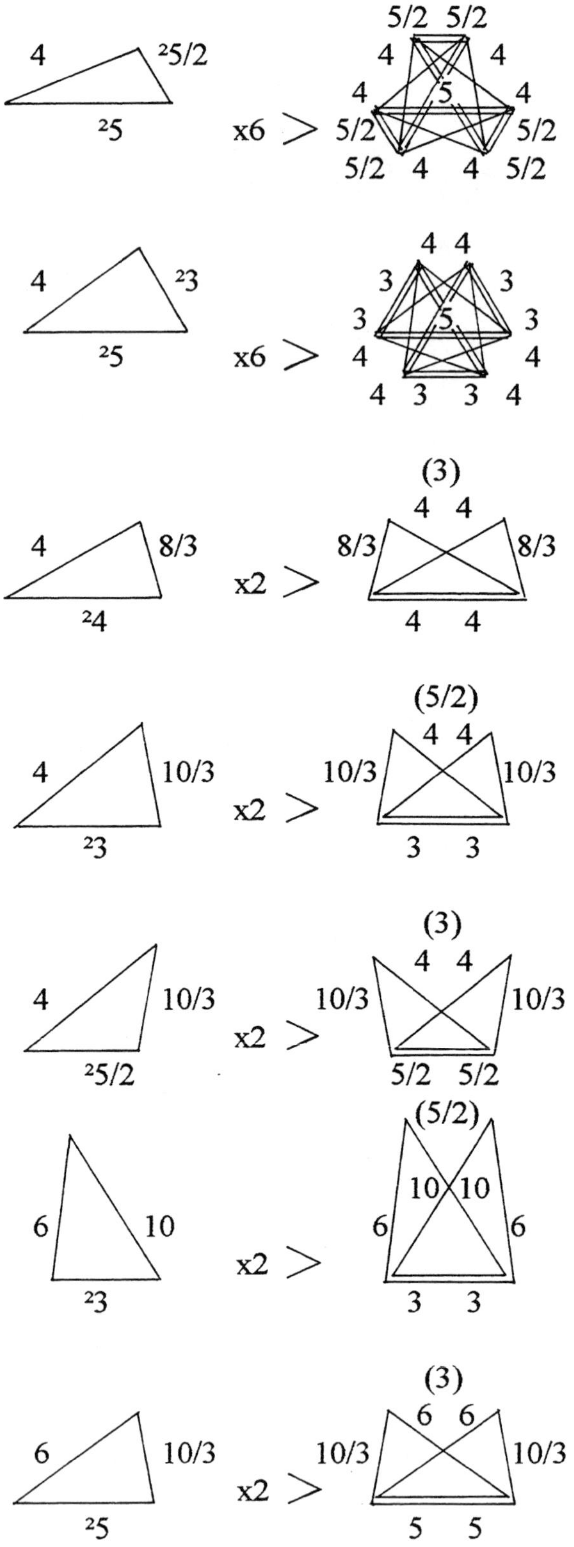

Fig.48: More Multiple Vertex Figures

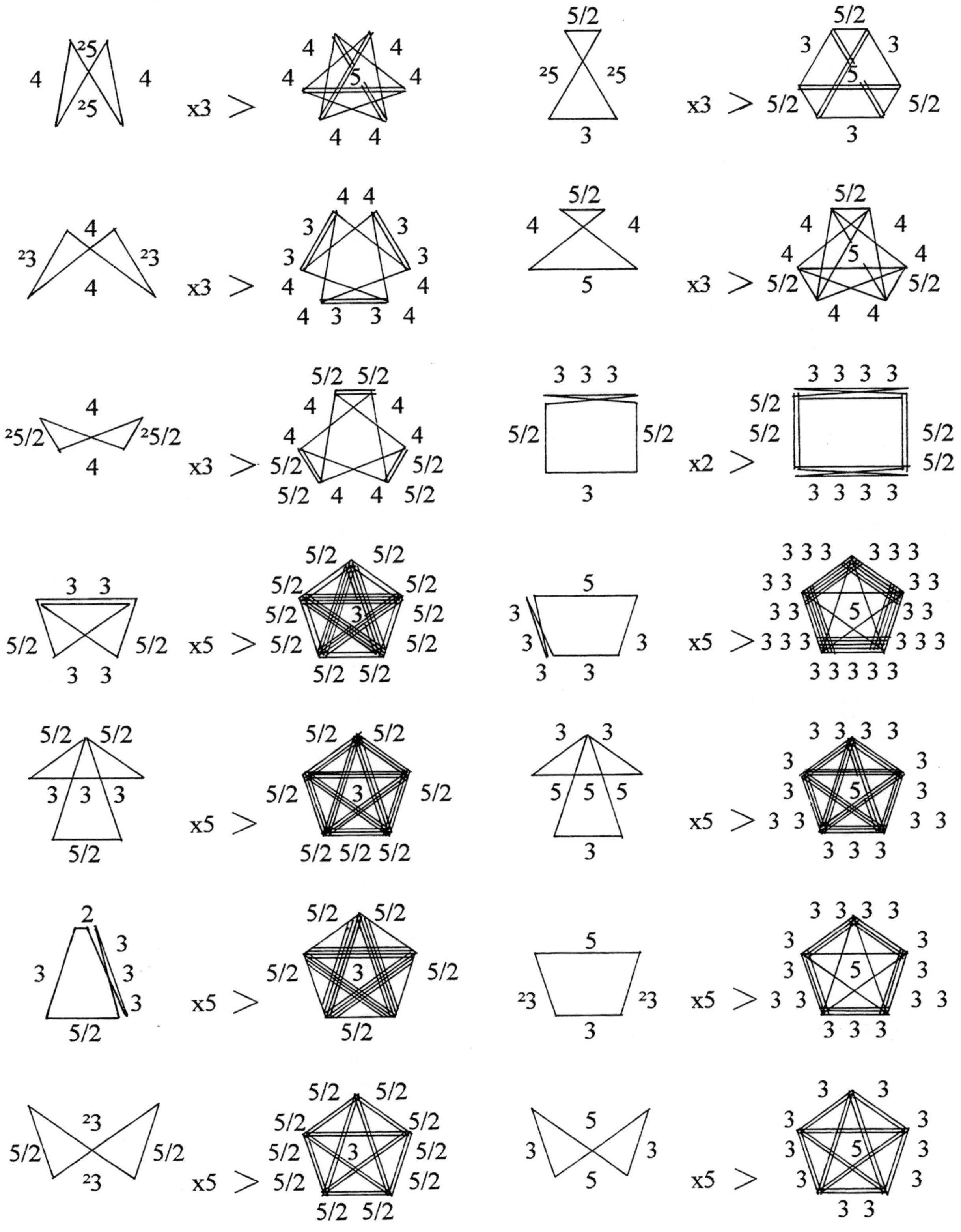
4 $^{2}5$ $^{2}5$ 4
x3 >
4 4 4 4 5 4 4
5/2 $^{2}5$ $^{2}5$ 3
x3 >
5/2 3 3 5 5/2 5/2 3
4 $^{2}3$ $^{2}3$ 4
x3 >
4 4 3 3 3 3 4 4 4 3 3 4
5/2 4 4 5
x3 >
5/2 4 4 4 5 4 5/2 5/2 4 4
4 $^{2}5/2$ $^{2}5/2$ 4
x3 >
5/2 5/2 4 4 4 4 5/2 5/2 5/2 4 4 5/2
3 3 3 5/2 5/2 3
x2 >
3 3 3 3 5/2 5/2 5/2 5/2 3 3 3 3
3 3 5/2 5/2 3 3
x5 >
5/2 5/2 5/2 5/2 5/2 5/2 5/2 3 5/2 5/2 5/2 5/2
5 3 3 3 3 3
x5 >
3 3 3 3 3 3 3 3 3 3 3 3 5 3 3 3 3 3 3 3 3 3 3 3 3
5/2 5/2 3 3 3 5/2
x5 >
5/2 5/2 5/2 3 5/2 5/2 5/2 5/2
3 3 5 5 5 3
x5 >
3 3 3 3 3 3 3 5 3 3 3 3 3 3 3 3
2 3 3 3 3 5/2
x5 >
5/2 5/2 5/2 3 5/2 5/2
5 $^{2}3$ $^{2}3$ 3
x5 >
3 3 3 3 3 3 3 5 3 3 3 3 3 3 3 3
$^{2}3$ 5/2 5/2 $^{2}3$
x5 >
5/2 5/2 5/2 5/2 5/2 3 5/2 5/2 5/2 5/2 5/2
5 3 3 5
x5 >
3 3 3 3 3 5 3 3 3 3 3

MONSTER POLYHEDRA

There remains one polyhedron shell that defies definition as part of a Schwarz triangle family. We have already seen it once as the shell that fits around the Great Snub Dodecicosidodecahedron on page 53.

The full shell consists of twenty interlocked octahedra joined by pentagram {5/2} faces. There are thus the eight triangles and four internal squares of two of these octahedra, plus two pentagrams, at each of the sixty vertices, as shown in Figure 49.

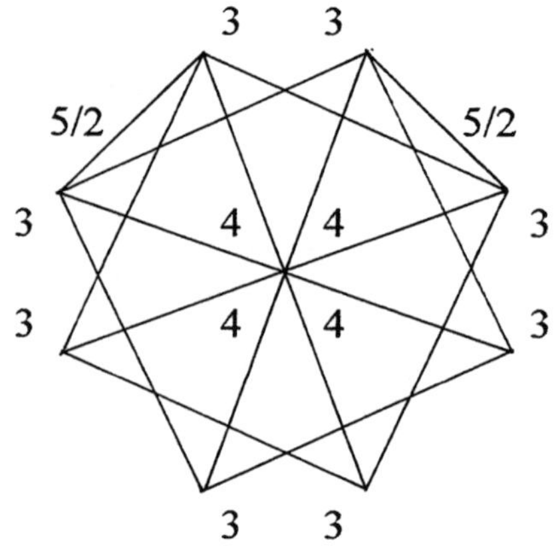

Fig.49: The Monster Vertex Figure

Missing out some of the faces of this monster shell and treating the squares as analogous to snub triangle faces there are two further polyhedra that can be counted amongst the 'uniform polyhedra'. One of these, the Great Dirhombicosidodecahedron, was described by Coxeter et al. as the 75th and last of the uniforms, being unusual in having eight faces per vertex. The other, the Great Disnub Dirhombidodecahedron, tops this with twelve, but is dismissed by purists on account of its having some double edges.

These both have four figure symbols and are shown adjoining in Figure 50 together with two further possibilities that are compounds of Octahedra or Tetrahemihexahedra.

The vertices of this shell are roughly those of a Rhombicosidodecahedron, in fact one that has been 'un-normalised' slightly so that the squares are rectangles and the triangle edges are longer than the pentagon edges. Although there is room for a pentagram at all four corners of the vertex figure, they can only fit in on two adjoining corners, which correspond to the two decagons within the Rhombicosidodecahedron's vertex figure. There is thus not a 5/2 5/3 |2, as might otherwise be expected.

<table>
<tr><th>Symbol</th><th>V</th><th>E</th><th>F</th></tr>
<tr><td>|3/2 5/3 3 5/2
92 (119)
Great Dirhombicosidodecahedron</td><td>60</td><td>240</td><td>40 △
60 □
24 ☆</td></tr>
<tr><td>|(3/2) 5/3 (3) 5/2
Skilling's Monster
Gt Disnub Dirhombidodecahedron</td><td>60</td><td>120x2
120x1</td><td>120 △
60 □
24 ☆</td></tr>
<tr><td>
Compound of
Twenty Tetrahemihexahedra</td><td>60x4</td><td>480</td><td>80 △
80 △
120 □</td></tr>
<tr><td>20{3,4}
Compound of
Twenty Octahedra</td><td>60x2</td><td>240</td><td>160 △</td></tr>
</table>

Table 25: Monster Polyhedra

|(3/2) 5/3 (3) 5/2
Skilling's Monster
Great Disnub
Dirhombidodecahedron

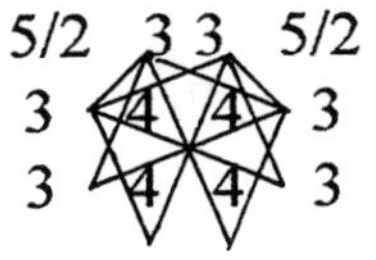

|3/2 5/3 3 5/2
92 (119)
Great
Dirhombicosidodecahedron

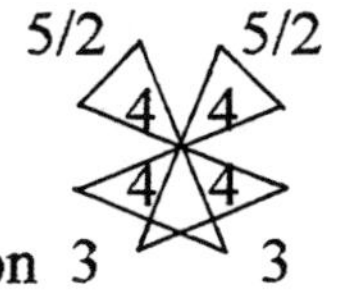

Compound of Twenty
Tetrahemihexahedra

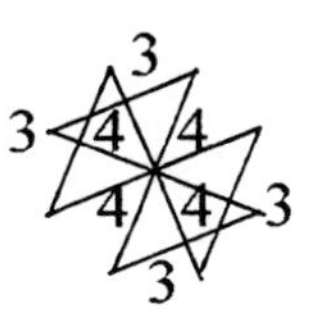

20{3,4}
Compound of
Twenty Octahedra

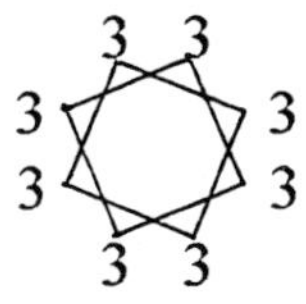

Fig.50: Monster Polyhedra

FIVE STAR SUB-FAMILIES

We have now explored thoroughly the five sub-families of the cross and star polyhedra. Their Schwarz triangles and associated colunar triangles are shown adjoining in Figures 50, 51, 52, 53 and 54.

In general we have found two types of 'quasi-regular' polyhedron in each sub-family: a double polyhedron and the quasi-regular polyhedron shown in Figure 40. There are three 'even faced' polyhedra in each sub-family: a double truncated parent polyhedron, its double quasitruncated equivalent and the quasiquasitruncated even faced star polyhedron shown in Figure 42.

In addition there are four 'rhombihedra' in each sub-family: a double quasi-regular polyhedron, the hemihedron shown in Figure 44, the rhombihedron shown in Figure 41 and the quasirhombihedron shown in Figure 43, all very much in parallel with what we saw for the regular polyhedra sub-families.

This might lead us to believe that there will be super-families formed by pairing up the sub-family Schwarz triangles. This however cannot be done, as it would involve a {4/6} or {5/6} occurring where the two triangles meet, so that we cannot form a new triangle.

We will however round things up now with an overview of all possible lunes and how they divide into colunar triangles, followed by a further brief appendix dealing with the star dihedra and prisms etc.

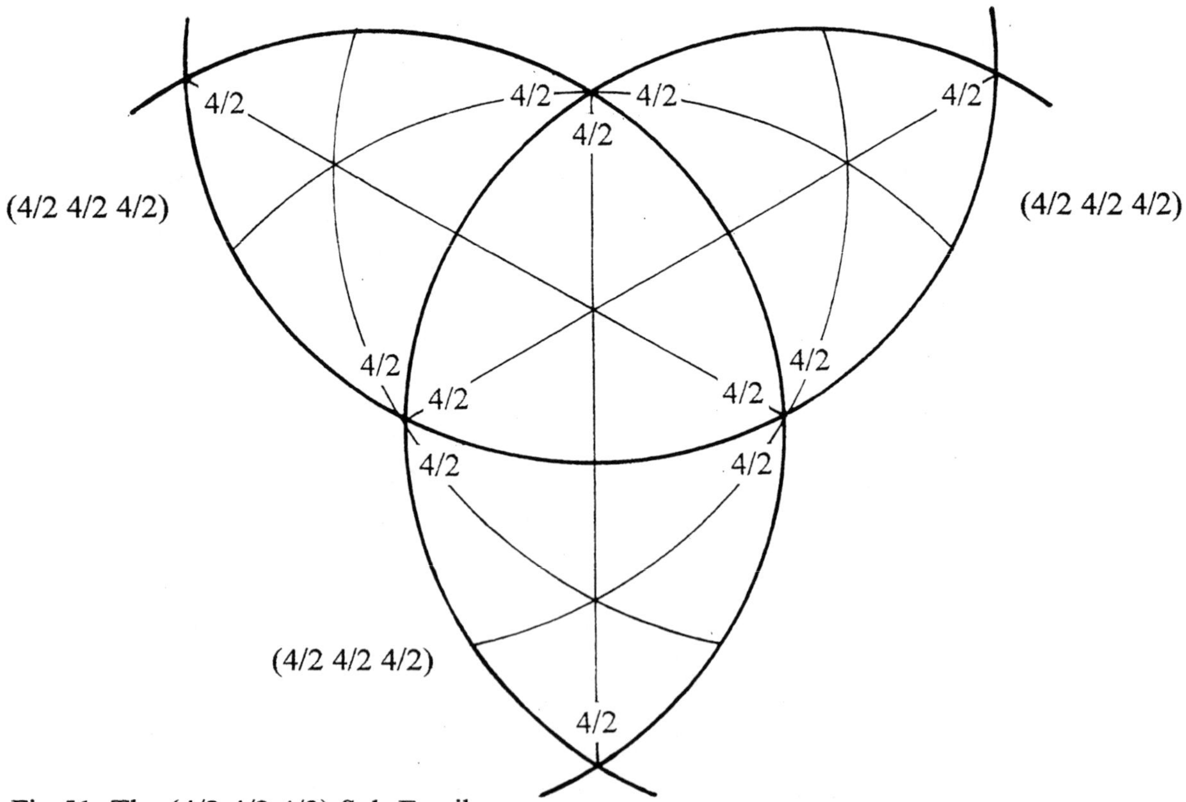

Fig.51: The (4/2 4/2 4/2) Sub-Family

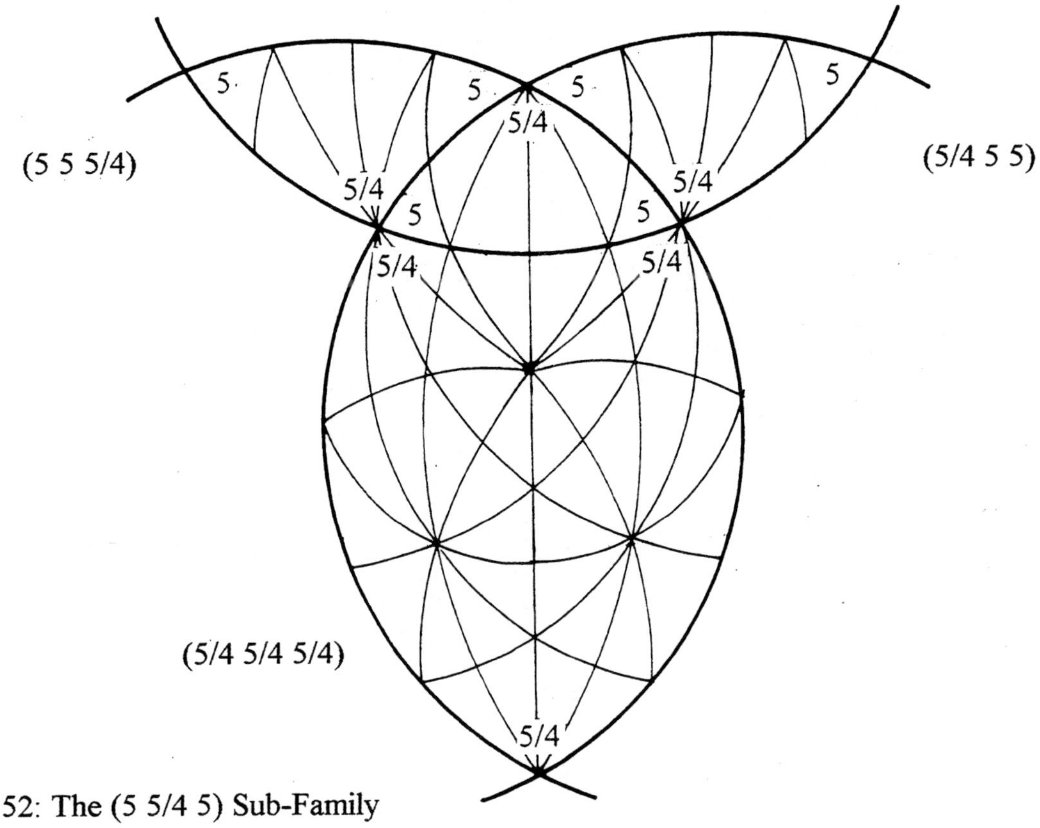

Fig.52: The (5 5/4 5) Sub-Family

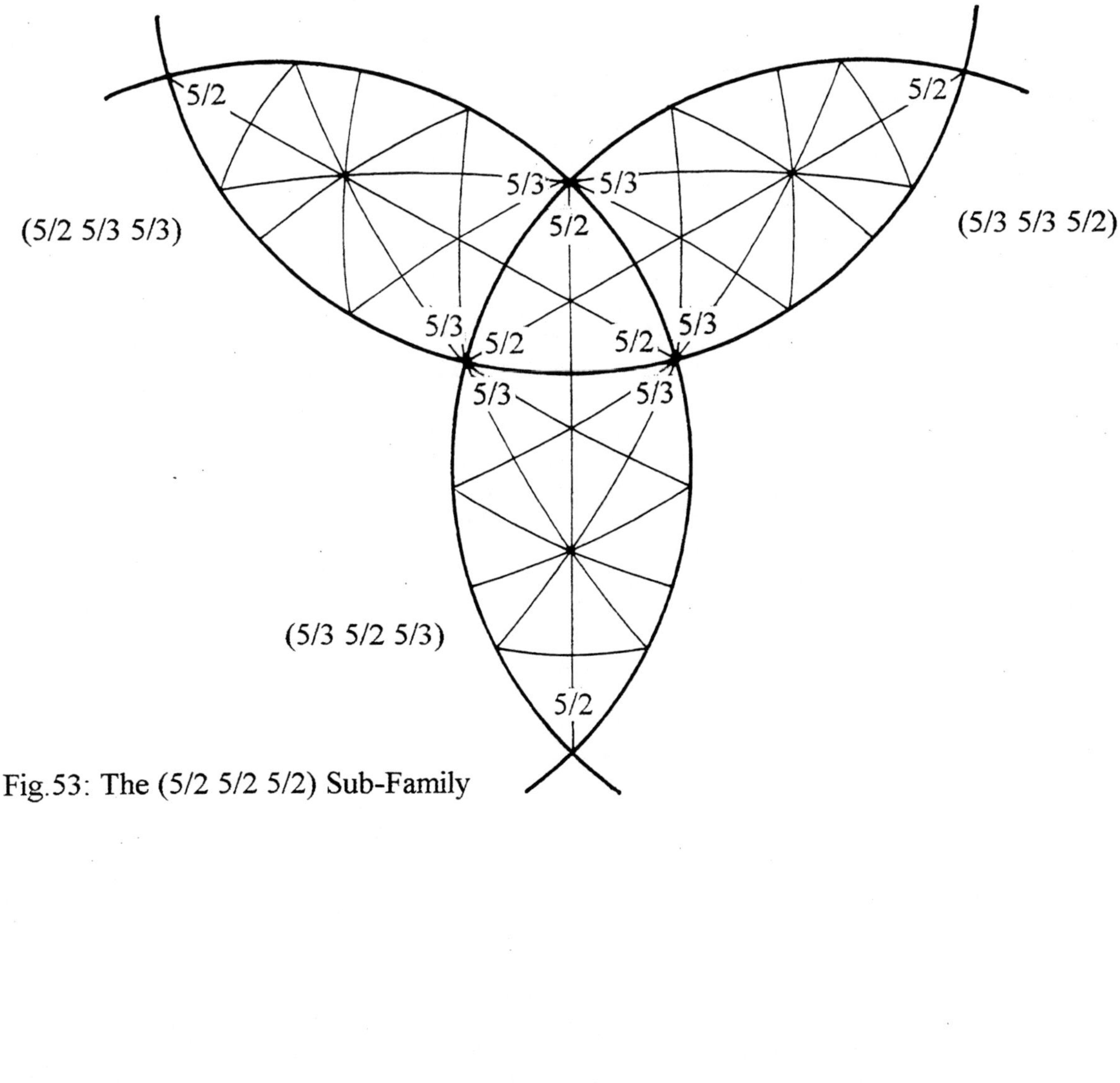

Fig.53: The (5/2 5/2 5/2) Sub-Family

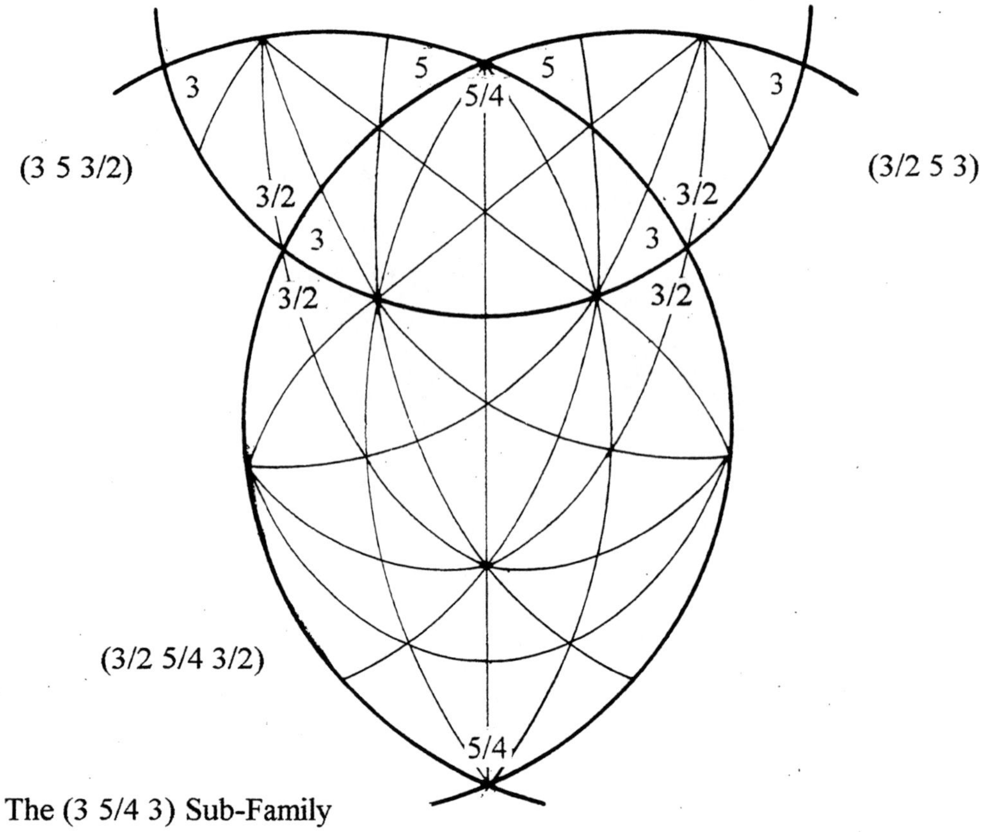

Fig.54: The (3 5/4 3) Sub-Family

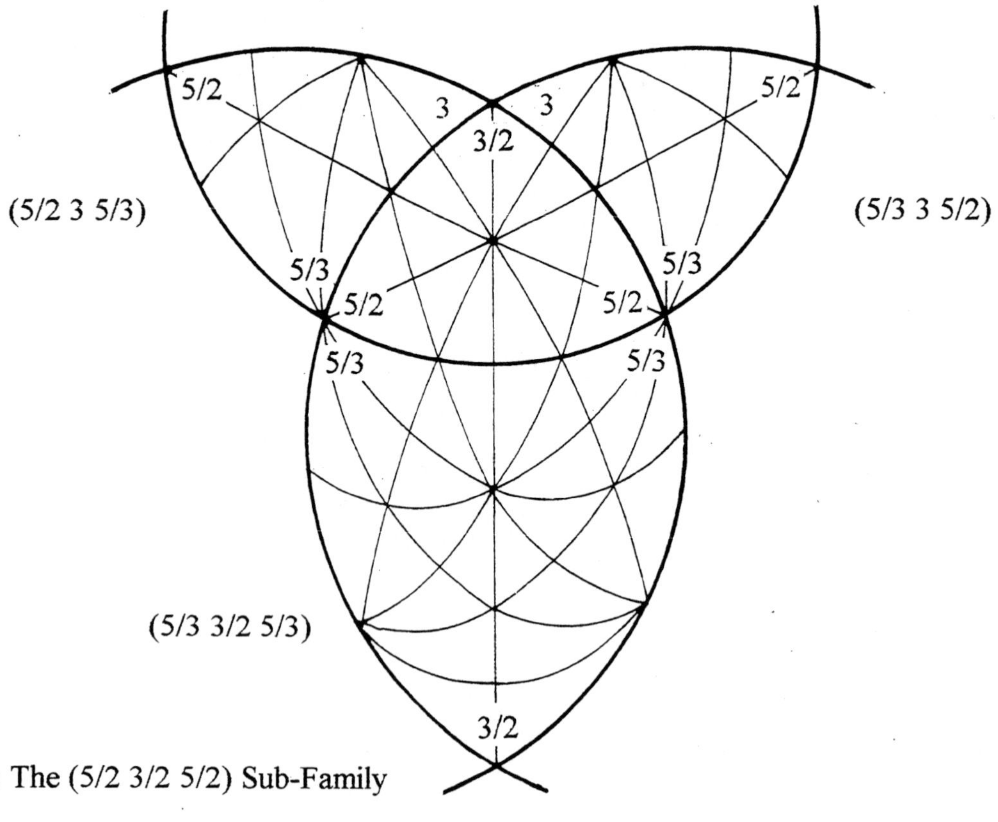

Fig.55: The (5/2 3/2 5/2) Sub-Family

ALL POSSIBLE SCHWARZ TRIANGLES

In Table 26 adjoining we summarise the various Schwarz triangles (x y z) we have examined so far, both in this volume and the previous one, '*The Simpler? Polyhedra*'.

These comprise the three regular polyhedra families, their five sub-families, the two super-families featured at the beginning of this volume, the three cross and star polyhedra families and five sub-families.

The corresponding colunar triangles to each Schwarz triangle are given alongside as (x y' z'), (x' y' z) and (x' y z'). We can find these same triangles and no others by examining the polyhedral kaleidoscopes and dividing each possible lune into two colunar triangles as shown in Figures 56, 57 and 58.

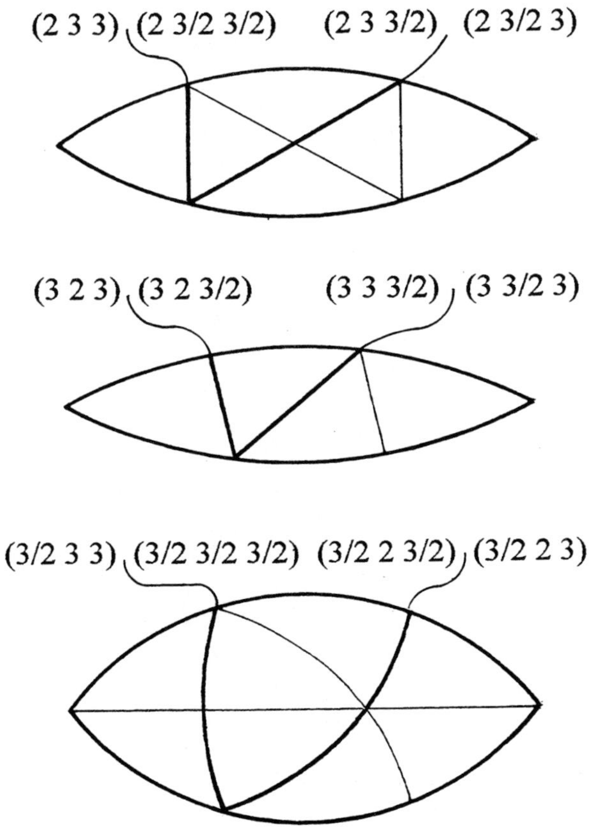

Fig.56: Tetrahedral Lunes

It thus appears that there are no further Schwarz triangles to be found and we have completed our survey of the polyhedra.

(x y z)	(x y' z')	(x' y' z)	(x' y z')
(3 2 3)	(3 2 3/2)	(3/2 2 3)	(3/2 2 3/2)
(3 2 4)	(3 2 4/3)	(3/2 2 4)	(3/2 2 4/3)
(3 2 5)	(3 2 5/4)	(3/2 2 5)	(3/2 2 5/4)
(3 3/2 3)	(3 3 3/2)	(3/2 3 3)	(3/2 3/2 3/2)
(4 3/2 4)	(4 3 4/3)	(4/3 3 4)	(4/3 3/2 4/3)
(5 3/2 5)	(5 3 5/4)	(5/4 3 5)	(5/4 3/2 5/4)
(3 4/2 3)	(3 4/2 3/2)	(3/2 4/2 3)	(3/2 4/2 3/2)
(3 5/2 3)	(3 5/3 3/2)	(3/2 5/3 3)	(3/2 5/2 3/2)
(3 4 4/3)	(3 4/3 4)	(3/2 4/3 4/3)	(3/2 4 4)
(3 5 5/3)	(3 5/4 5/2)	(3/2 5/4 5/3)	(3/2 5 5/2)
(4 2 4/2)	(4 2 4/2)	(4/3 2 4/2)	(4/3 2 4/2)
(5 2 5/2)	(5 2 5/3)	(5/4 2 5/2)	(5/4 2 5/3)
(3 2 5/2)	(3 2 5/3)	(3/2 2 5/2)	(3/2 2 5/3)
(4/2 4/2 4/2)	(4/2 4/2 4/2)	(4/2 4/2 4/2)	(4/2 4/2 4/2)
(5/2 5/2 5/2)	(5/2 5/3 5/3)	(5/3 5/3 5/2)	(5/3 5/2 5/3)
(5/2 3/2 5/2)	(5/2 3 5/3)	(5/3 3 5/2)	(5/3 3/2 5/3)
(5 5/4 5)	(5 5 5/4)	(5/4 5 5)	(5/4 5/4 5/4)
(3 5/4 3)	(3 5 3/2)	(3/2 5 3)	(3/2 5/4 3/2)

Table 26: Colunar Triangles

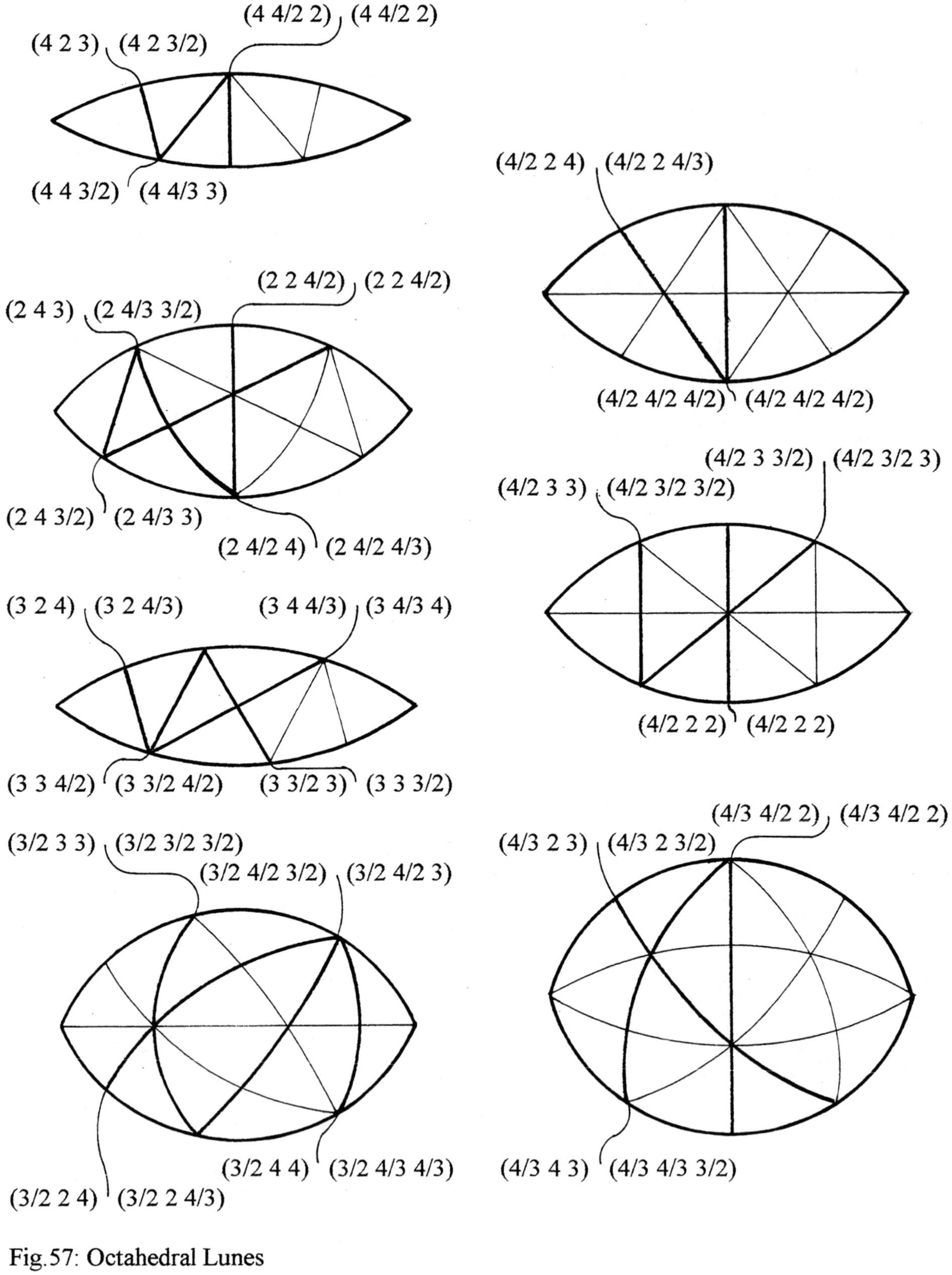

Fig.57: Octahedral Lunes

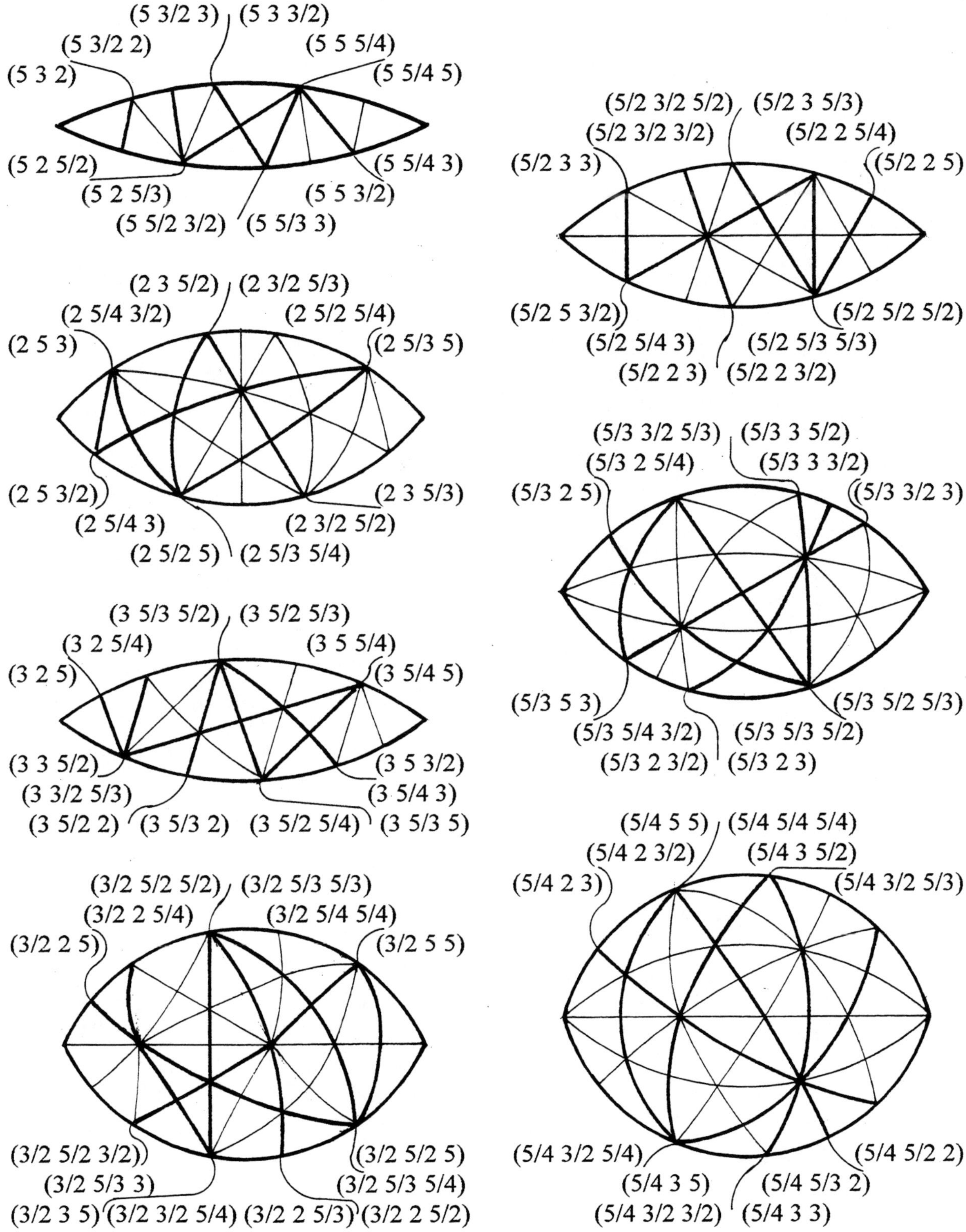

Fig.58: Icosahedral Lunes

STAR DIHEDRA

In the last volume we considered the case of regular polygons placed back to back, two per vertex, and produced the interesting degenerate polyhedra known as the dihedra. We will now examine this proposition more generally for any cross or star polygon {n/d}. The result that we get is two cross or star polygons back to back: 2| 2 n/d.

As before these dihedra have length and sometimes area but no volume, a property that excludes them from full membership of the uniform polyhedra. They do, however, have n vertices, n edges and 2 faces and thus still obey Euler's formula: V+F=E+2, or more generally V+F=E+2d, if we account for multiple density faces d times. Five members of this infinite set are shown in Figure 59 and described in Table 27 below.

Symbol		V	E	F
2| 2 3/2	{3/2,2}	3	3	2 △
Double Triangle				
2| 2 4/2	{4/2,2}	4	4	2 ✕
Double Cross				
2| 2 5/2	{5/2,2}	5	5	2 ☆
Double Pentagram				
2| 2 6/2	{6/2,2}	6	6	2 ✡
Double Hexagram				
2| 2 6/3	{6/3,2}	6	6	2 ✳
Double Hexacross				

Table 27: Cross & Star Dihedra

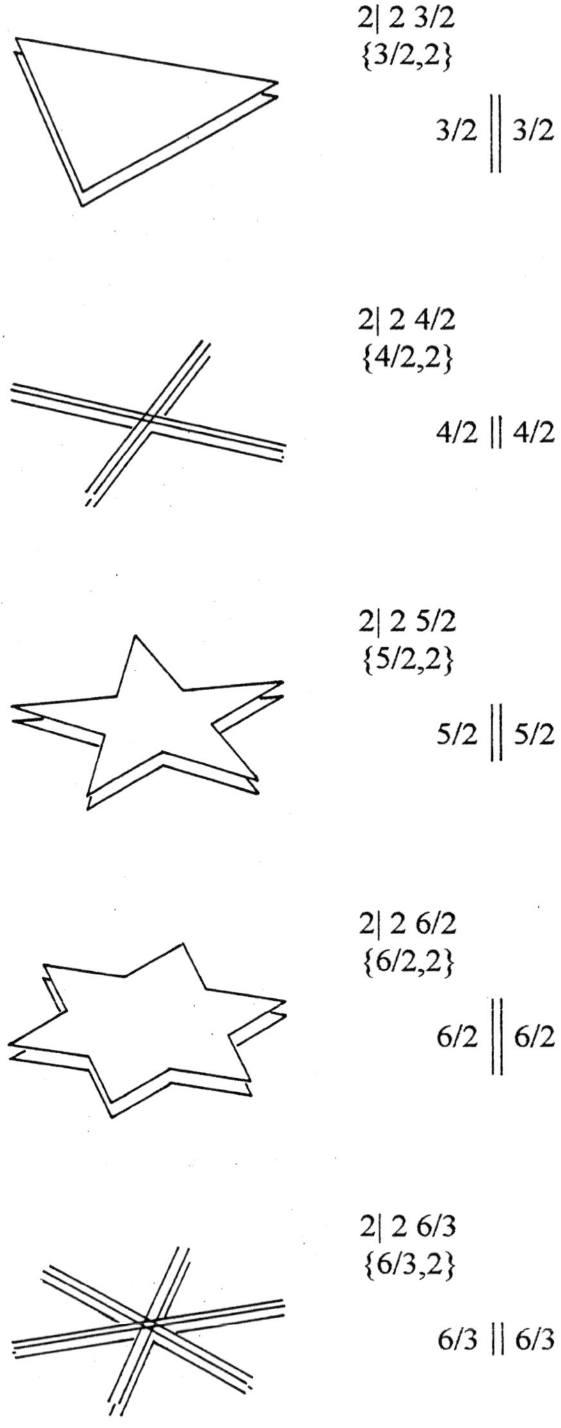

Fig.59: Cross & Star Dihedra

MULTIPLE STAR DIGONS

The cross and star dihedra can be found to have duals in the same way that the regular dihedra were in the last volume.

These are further regular polyhedra at the other end of a truncation series, i.e. {2,n/d}, consisting of multiple bundles of digon faces, meeting at two opposing vertices.

These multiple digons have the general symbol n/d| 2 2 and are shown in Figure 60 and described in Table 28 below.

Again they can be more clearly appreciated in terms of a dihedral kaleidoscope of (2 2 n/d) Möbius triangles, which necessarily wrap around the sphere d times to close.

Symbol		V	E	F	
3/2| 2 2	{2,3/2}	2	3	3	||
Triple Digon					
4/2| 2 2	{2,4/2}	2	4	4	||
Quadruple Digon					
5/2| 2 2	{2,5/2}	2	5	5	||
Quintuple Digon					
6/2| 2 2	{2,6/2}	2	6	6	||
Sextuple Digon					
6/3| 2 2	{2,6/3}	2	6	6	||
Sextuple Digon					

Table 28: Multiple Digons

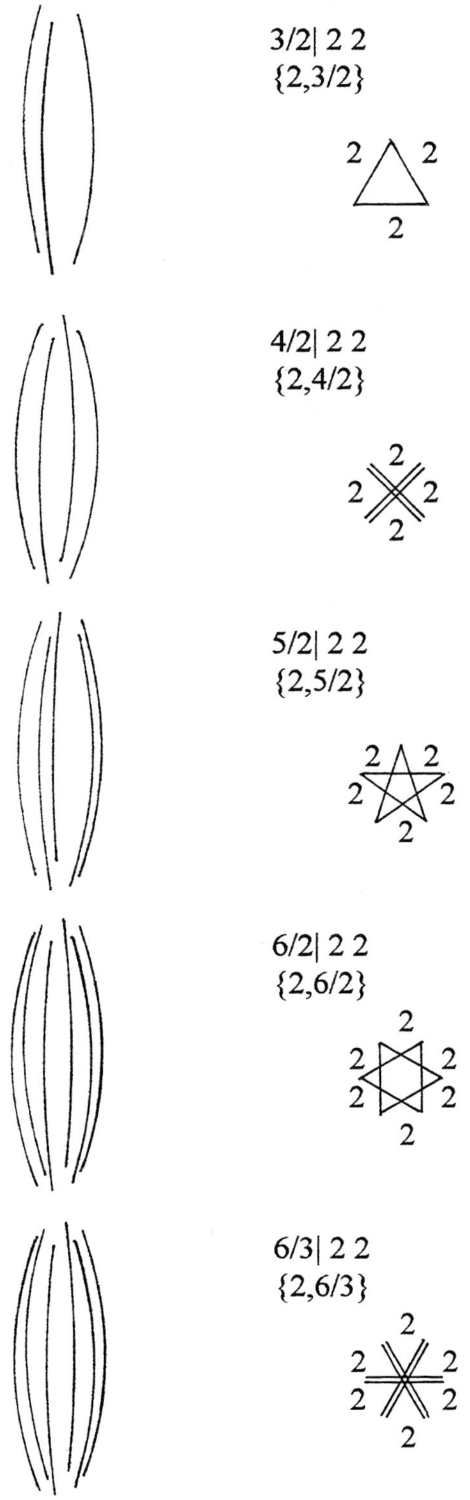

Fig.60: Multiple Digons

TRUNCATED DIGONS: STAR PRISMS

The results of truncating the multiple star digons we have just examined are good justification for including such primitive and degenerate assemblies of polygons.

The theoretical truncation of a digon {2} face is a square {4}, so that what we get as t{2,n/d} are the cross and star prisms consisting of two {n/d} faces and n squares each: n/d 2 |2.

These are shown in Figure 61 and described in Table 29 below, and can be seen to have the usual isosceles triangle vertex figure of a truncated polyhedron . In fact some of these are the first true polyhedra we have seen in this section as they have finite volumes.

Symbol		V	E	F
3/2 2 |2 Triangle Prism	t{2,3/2}	6	9	3 □ 2 △
4/2 2 |2 Cross Prism	t{2,4/2}	8	12	4 □ 2 ✕
5/2 2 |2 Pentagram Prism	t{2,5/2}	10	15	5 □ 2 ☆
6/2 2 |2 Hexagram Prism	t{2,6/2}	12	18	6 □ 2 ✡
6/3 2 |2 Hexacross Prism	t{2,6/3}	12	18	6 □ 2 ✳

Table 29: Truncated Digons

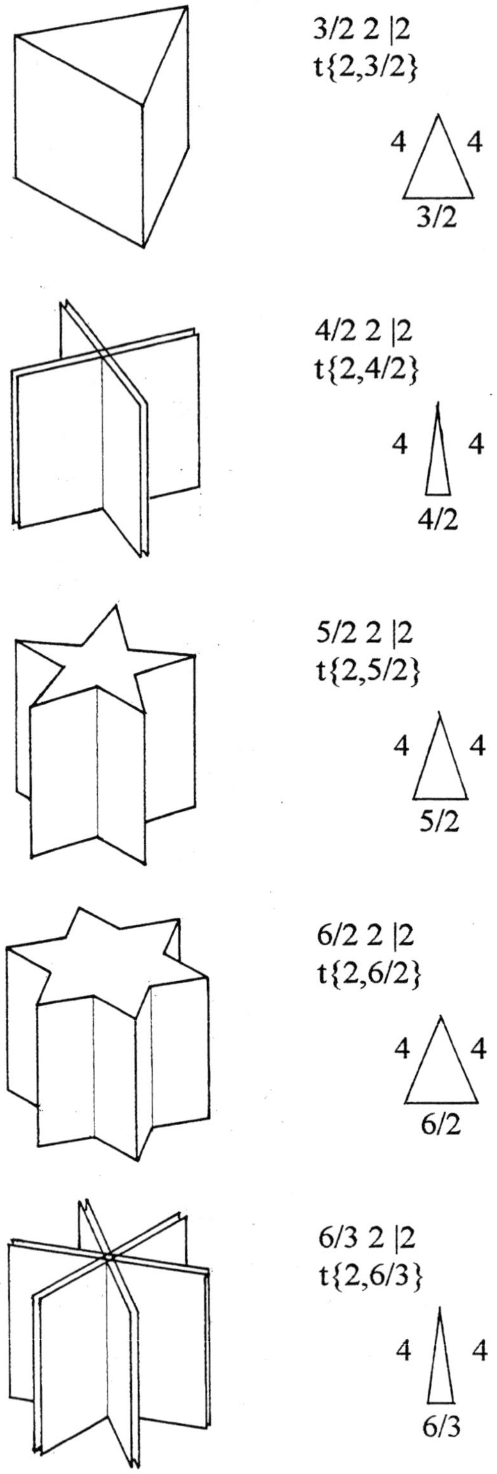

Fig.61: Truncated Digons

TRUNCATED STAR DIHEDRA

Truncating the other member of our dual pairs, the cross and star dihedra we started with, we get further star dihedra t{n/d,2}, consisting of two truncated polygons t{n/d} back to back, with every other edge made double by the inclusion of a digon face there: 2 2 |n/d.

When the truncated polygons are themselves double we get quadruple polygons resulting, as described in Table 30 below and shown in Figure 62 adjoining.

Again these can be seen to have the usual truncated polyhedron isosceles triangle vertex figure, although very slender on account of the digon {2} face with its dual edge length of zero, shown here with a minimal length for clarity.

Symbol		V	E	F
2 2 |3/2	t{3/2,2}	3x2	3x3	2 △ 3 ||
Quadruple Triangle				
2 2 |4/2	t{4/2,2}	4x2	4x3	2 □ 4 ||
Quadruple Square				
2 2 |5/2	t{5/2,2}	5x2	5x3	2 ⬠ 5 ||
Quadruple Pentagon				
2 2 |6/2	t{6/2,2}	6x2	6x3	2 ⬡ 6 ||
Quadruple Hexagon				
2 2 |6/3	t{6/3,2}	12	6x2 6x1	2 ✡ 6 ||
Double Dodecagram				

Table 30: Truncated Cross & Star Dihedra

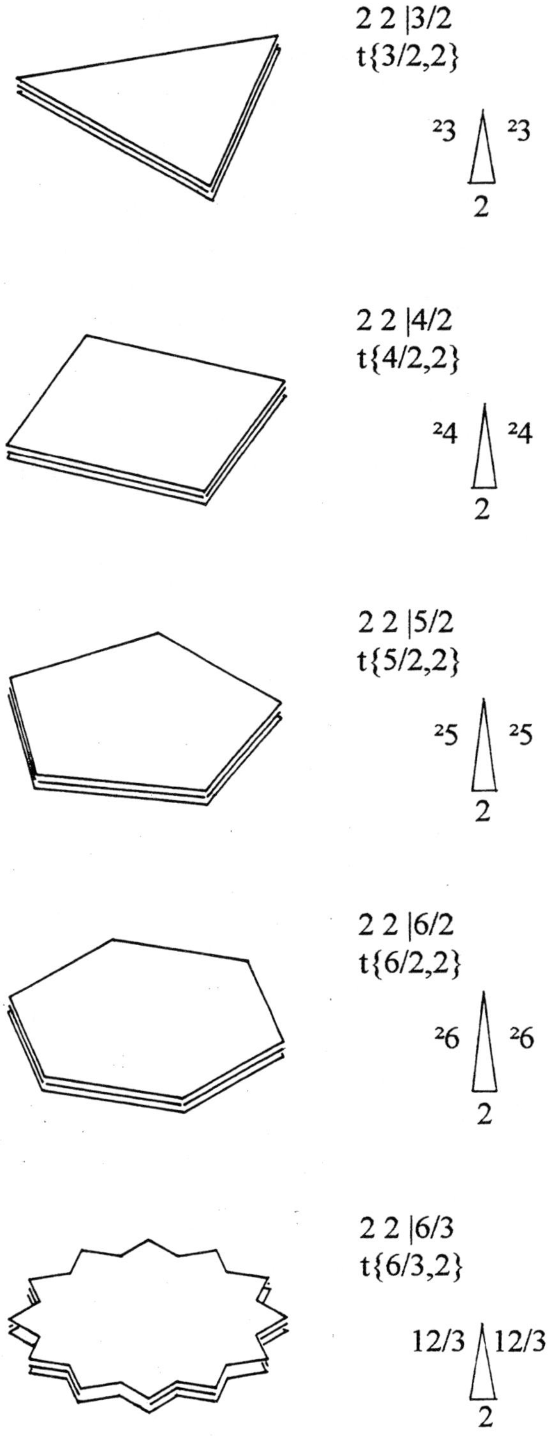

Fig.62: Truncated Cross & Star Dihedra

QUASI-REGULAR STAR DIHEDRA

Continuing the truncation process onwards from either of our truncated forms, we can obtain quasi-regular cross or star dihedra at the centre of each truncation series: 2| n/d 2, the same as the parent cross or star dihedra.

These cross and star dihedra again have length and some have area but no volume, consisting of two parent cross or star polygons back to back, but now joined with n double edges all round, containing the n extra digon faces.

Here if we again show the digon {2} faces with a minimal rather than zero dual edge length, we can construct meaningful rectangular vertex figures, as shown in Figure 63. These quasi-regular star dihedra are further described in Table 31 below.

Symbol		V	E	F
2| 3/2 2 Double Triangle	{2 / 3/2}	3	3x2	2 △ 3 ‖
2| 4/2 2 Double Cross	{2 / 4/2}	4	4x2	2 ✕ 4 ‖
2| 5/2 2 Double Pentagram	{2 / 5/2}	5	5x2	2 ☆ 5 ‖
2| 6/2 2 Double Hexagram	{2 / 6/2}	2	6x2	2 ✡ 6 ‖
2| 6/3 2 Double Hexacross	{2 / 6/3}	6	6x2	2 ✳ 6 ‖

Table 31: Quasi-Regular Cross & Star Dihedra

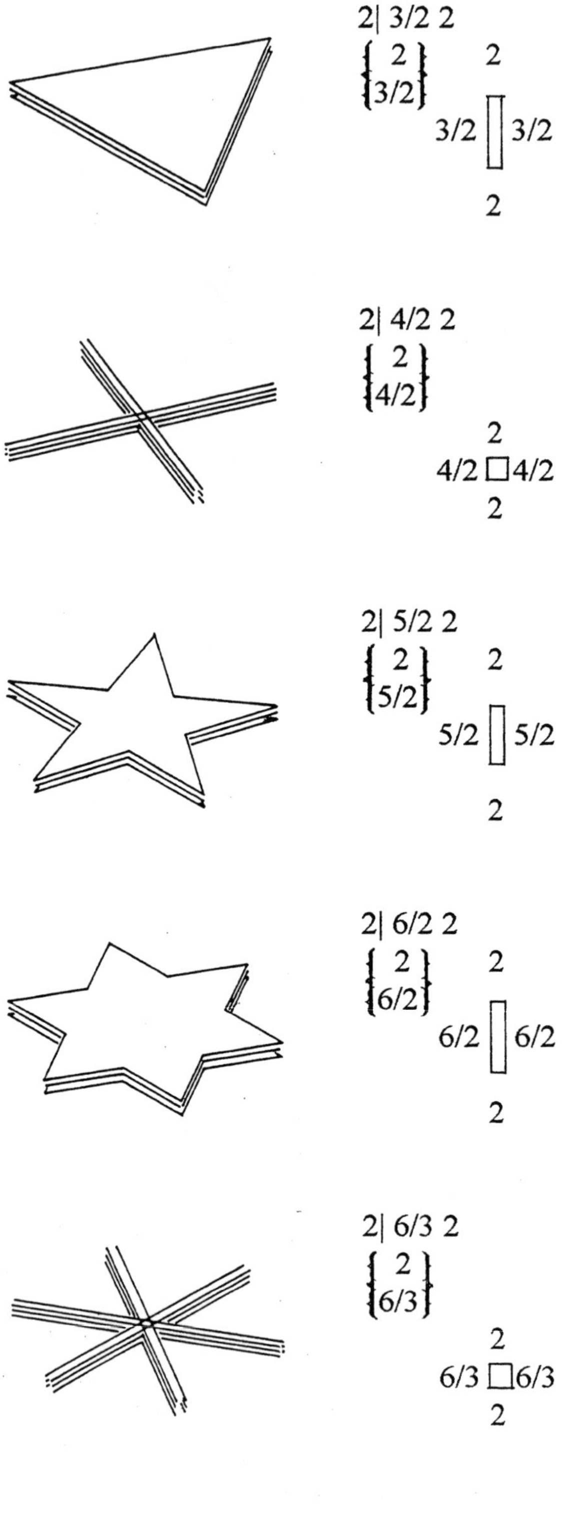

Fig.63: Quasi-Regular Cross & Star Dihedra

EVEN FACED STAR DIHEDRA

We have now seen dihedral symmetry family equivalents of all members of the normal polyhedral symmetry families' first truncation series.

Taking the quasi-regular cross and star dihedra we can produce a first truncation in a second series which will be equivalent to the even faced polyhedra: 2 2 n/d|.

These even faced star dihedra are shown in Figure 64 and described in Table 32 below.

They consist of prisms with two truncated polygon faces t{n/d} plus 2n square faces, alternately truncated digons and new truncating faces, and have the usual even faced polyhedron triangular vertex figure.

Symbol		V	E	F
2 2 3/2| Double Triangle Prism	t{2, 3/2}	6x2	6x2 6x1	2 △ 3 □ 3 □
2 2 4/2| 2{4,3} Double Square Prism	t{2, 4/2}	8x2	8x2 8x1	2 □ 4 □ 4 □
2 2 5/2| Double Pentagon Prism	t{2, 5/2}	10x2	10x2 10x1	2 ⬠ 5 □ 5 □
2 2 6/2| Double Hexagon Prism	t{2, 6/2}	12x2	12x2 12x1	2 ⬡ 6 □ 6 □
2 2 6/3| 3{4,3} Dodecagram Prism	t{2, 6/3}	24	36	2 ✡ 6 □ 6 □

Table 32: Even Faced Cross & Star Dihedra

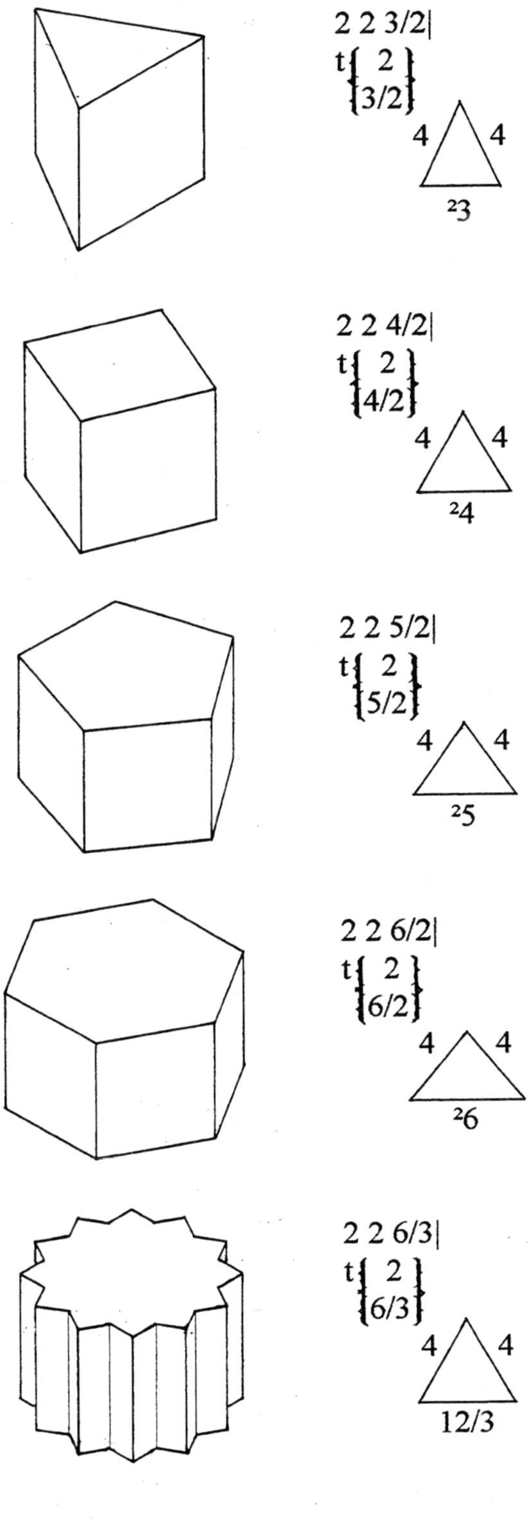

Fig.64: Even Faced Cross & Star Dihedra

STAR RHOMBIDIHEDRA

Continuing this second series truncation process leads ultimately to a dual consisting of n rhomb faces, n of which meet at two vertices and two of which meet at n vertices; essentially another multiple set of bunched digons (where each digon is an elongated rhomb), but this time with extra vertices at the equatorial positions.

En route to this we can find a second member of the second series, the equivalent of the rhombihedra, consisting of two {n/d} faces, n square faces and n double edged digon faces: 2 n/d |2. These have the usual rhombihedron isosceles trapezoid vertex figure consisting of two squares joining together {n/d} and a digon {2} as shown in Figure 65 and described in Table 33.

Symbol		V	E	F
2 3/2 |2	$r\begin{Bmatrix}2\\3/2\end{Bmatrix}$	6	3x2 6x1	2 △ 3 ‖ 3 □
Triangle Prism				
2 4/2 |2	$r\begin{Bmatrix}2\\4/2\end{Bmatrix}$	8	4x2 8x1	2 ✕ 4 ‖ 4 □
Cross Prism				
2 5/2 |2	$r\begin{Bmatrix}2\\5/2\end{Bmatrix}$	10	5x2 10x1	2 ☆ 5 ‖ 5 □
Pentagram Prism				
2 6/2 |2	$r\begin{Bmatrix}2\\6/2\end{Bmatrix}$	12	6x2 12x1	2 ✡ 6 ‖ 6 □
Hexagram Prism				
2 6/3 |2	$r\begin{Bmatrix}2\\6/3\end{Bmatrix}$	12	6x2 12x1	2 ✳ 6 ‖ 6 □
Hexacross Prism				

Table 33: Cross & Star Rhombidihedra

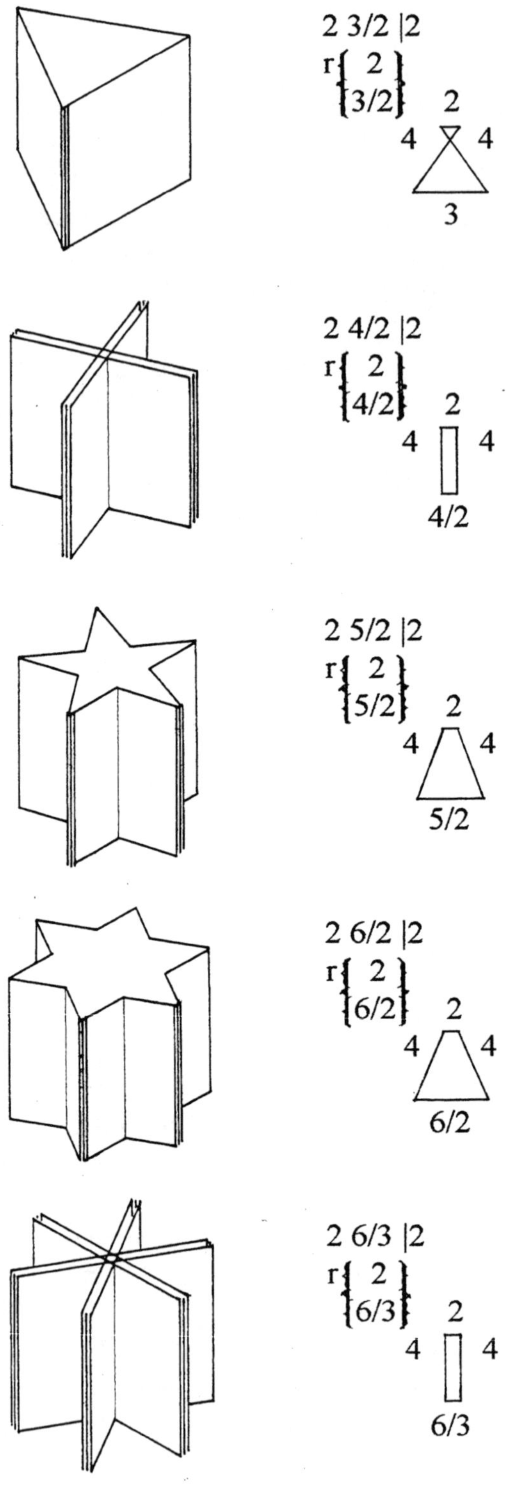

Fig.65: Cross & Star Rhombidihedra

SNUB STAR DIHEDRA: ANTIPRISMS

Taking the cross and star rhombidihedra we have just examined, we can remove the square faces, apply the characteristic snub twist and replace them each with two triangles to get snub cross and star dihedra to complete our second series.

These have two {n/d} faces, 2n triangle faces and n double edged digon faces and are known as antiprisms: |2 2 n/d. They have the usual irregular pentagon vertex figure of snubs and are shown in Figure 66 and described below in Table 34.

The first of this infinite set is interesting as it folds down into an eightfold triangle whilst the others include the Stella Octangula (two tetrahedra), a compound of two octahedra and a compound of three tetrahedra.

<table>
<tr><th>Symbol</th><th>V</th><th>E</th><th>F</th></tr>
<tr><td>|2 2 3/2 s{2 / 3/2}
Eightfold Triangle</td><td>6</td><td>3x2
9x1</td><td>2 △
3 ||
6 △</td></tr>
<tr><td>|2 2 4/2 s{2 / 4/2}
2{3,3}
Cross Antiprism (Stella Octangula)</td><td>8</td><td>4x2
12x1</td><td>2 ✕
4 ||
8 △</td></tr>
<tr><td>|2 2 5/2 s{2 / 5/2}
Pentagram Antiprism</td><td>10</td><td>5x2
15x1</td><td>2 ☆
5 ||
10 △</td></tr>
<tr><td>|2 2 6/2 s{2 / 6/2}
2{3,4}
Hexagram Antiprism</td><td>12</td><td>6x2
18x1</td><td>2 ✡
6 ||
12 △</td></tr>
<tr><td>|2 2 6/3 s{2 / 6/3}
3{3,3}
Hexacross Antiprism</td><td>12</td><td>6x2
18x1</td><td>2 ✳
6 ||
12 △</td></tr>
</table>

Table 34: Snub Cross & Star Dihedra

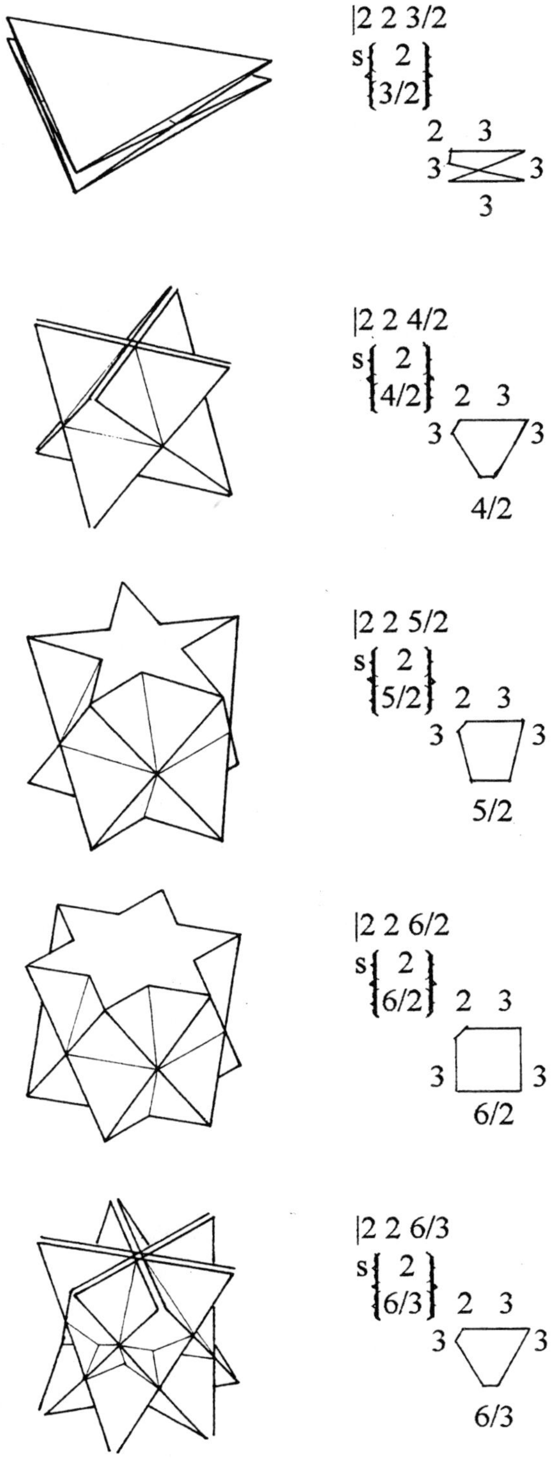

Fig.66: Snub Cross & Star Dihedra

RETROSNUB STAR DIHEDRA

Because the {2} face of a rhombidihedron has zero edge length as part of the vertex figure, it will appear the same normal or twisted. This means there will be no star quasirhombidihedra and consequently no quasisnub star dihedra.

It is however possible to derive a second snub star dihedron in certain cases, those of general form |2 2 k', when k = n/d, the value of the polygon, lies between 2 and 3, i.e 2>k'>3/2.

The first three examples of these retrosnub star dihedra, known as crossed antiprisms, are shown below in Figure 67, with above them for comparison two further members of the normal antiprism snub star dihedra.

<table>
<tr><th>Symbol</th><th></th><th>V</th><th>E</th><th>F</th></tr>
<tr><td>|2 2 7/3
Heptagram Antiprism</td><td>$s\begin{Bmatrix}2\\7/3\end{Bmatrix}$</td><td>14</td><td>7x2
21x1</td><td>2
7 ||
14 △</td></tr>
<tr><td>|2 2 8/3
Octagram Antiprism</td><td>$s\begin{Bmatrix}2\\8/3\end{Bmatrix}$</td><td>16</td><td>8x2
24x1</td><td>2
8 ||
16 △</td></tr>
<tr><td>|2 2 5/3
Pentagram Crossed Antiprism</td><td>$s\begin{Bmatrix}2\\5/3\end{Bmatrix}$</td><td>10</td><td>5x2
15x1</td><td>2
5 ||
10 △</td></tr>
<tr><td>|2 2 7/4
Heptagram Crossed Antiprism</td><td>$s\begin{Bmatrix}2\\7/4\end{Bmatrix}$</td><td>14</td><td>7x2
21x1</td><td>2
7 ||
14 △</td></tr>
<tr><td>|2 2 8/5
Octagram Crossed Antiprism</td><td>$s\begin{Bmatrix}2\\8/5\end{Bmatrix}$</td><td>16</td><td>8x2
24x1</td><td>2
8 ||
16 △</td></tr>
</table>

Table 35: Snub & Retrosnub Star Dihedra

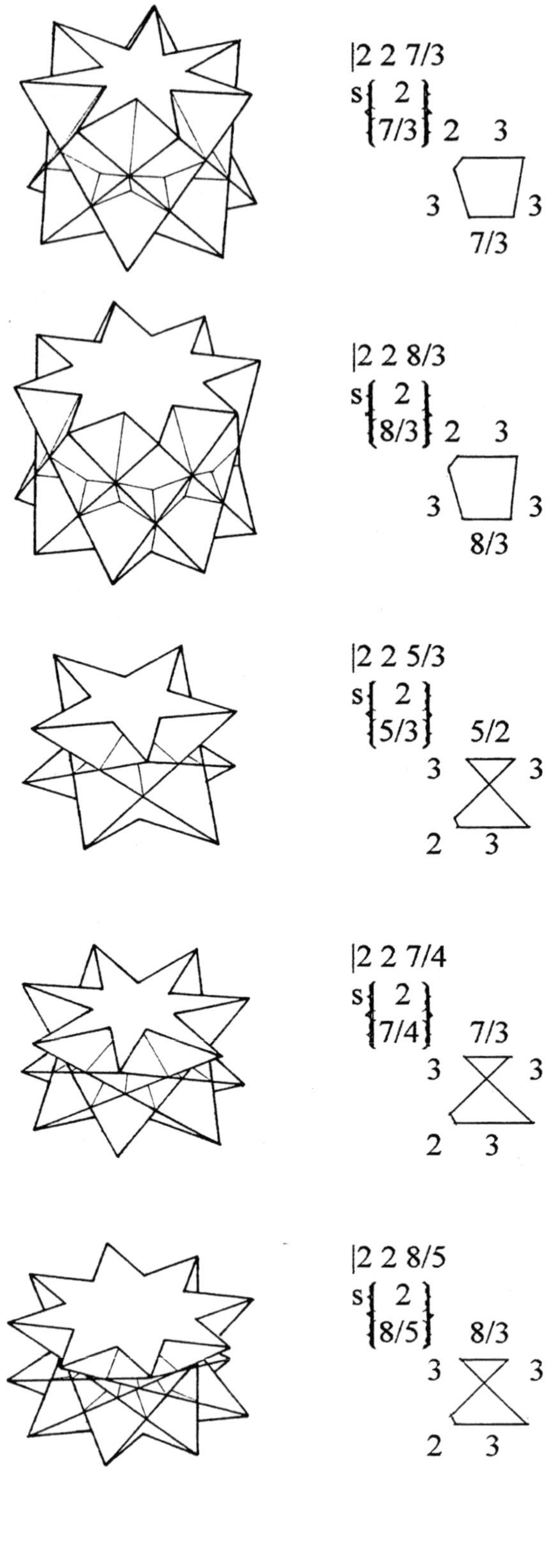

Fig.67: Snub & Retrosnub Star Dihedra

QUASITRUNCATED STAR DIHEDRA

An interesting property of the antiprisms is that halfway down each, a section taken will reveal the truncation of the star polygon on the upper or lower face. This property is useful for visualising quasitruncation, as when the star polygon face is technically an inverse and the section is taken through the crossed antiprism, the quasitruncated star polygon results.

We can quasitruncate the star dihedra themselves to produce t{n/d',2}, consisting of two t{n/d'} faces back to back, joined together by n double edges including digons, as shown in Figure 68 and described in Table 36 below: 2 2 |n/d'. Where the truncations are double polygons, once again we get quadruple results.

Symbol		V	E	F
2 2 |3/1 Double Hexagon	t{3/2',2}	6	3x2 3x1	2 3 ||
2 2 |4/2 Quadruple Square	t{4/2',2}	4x2	4x2 4x1	2 4 ||
2 2 |5/3 Double Decagram	t{5/2',2}	10	5x2 5x1	2 5 ||
2 2 |6/4 Quadruple Hexagram	t{6/2',2}	6x2	6x2 6x1	2 6 ||
2 2 |6/3 Double Dodecagram	t{6/3',2}	12	6x2 6x1	2 6 ||

Table 36: Quasitruncated Star Dihedra

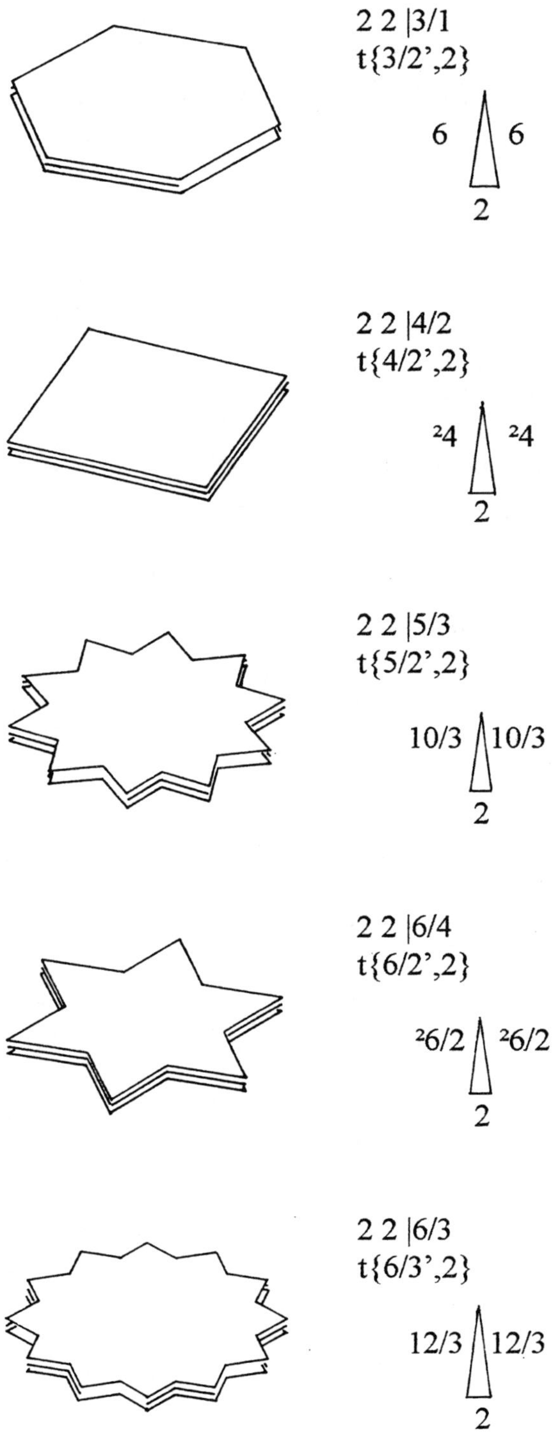

Fig.68: Quasitruncated Star Dihedra

QUASI EVEN FACED STAR DIHEDRA

The second quasitruncation we can consider is that of the quasi-regular star dihedra to produce quasi even faced star dihedra, prismatic forms of the quasi-truncated star dihedra just seen on the previous page.

They have similar triangular vertex figures to the even faced star dihedra, but with two squares and one quasitruncated {n/d} at each vertex: 2 2 n/d'|. These quasi even faced dihedra are shown in Figure 69 and described in Table 37 below.

We have now covered the simplest star and cross dihedral families, all those with n < 7; there would never be enough space for them all as the set of polygons is infinite, so we will stop here whilst we are ahead.

Symbol		V	E	F
2 2 3/1|	t{ 2 / 3/2' }	12	18	2 ⬡
				3 □
Hexagon Prism				3 □
2 2 4/2|	t{ 2 / 4/2' }	8x2	8x2	2 ⧉
2{4,3}			8x1	4 □
Double Square Prism				4 □
2 2 5/3|	t{ 2 / 5/2' }	20	30	2 ☆
				5 □
Decagram Prism				5 □
2 2 6/4|	t{ 2 / 6/2' }	12x2	12x2	2 ✡
			12x1	6 □
Double Hexagram Prism				6 □
2 2 6/3|	t{ 2 / 6/3' }	24	36	2 ☆
3{4,3}				6 □
Dodecagram Prism				6 □

Table 37: Quasi Even Faced Star Dihedra

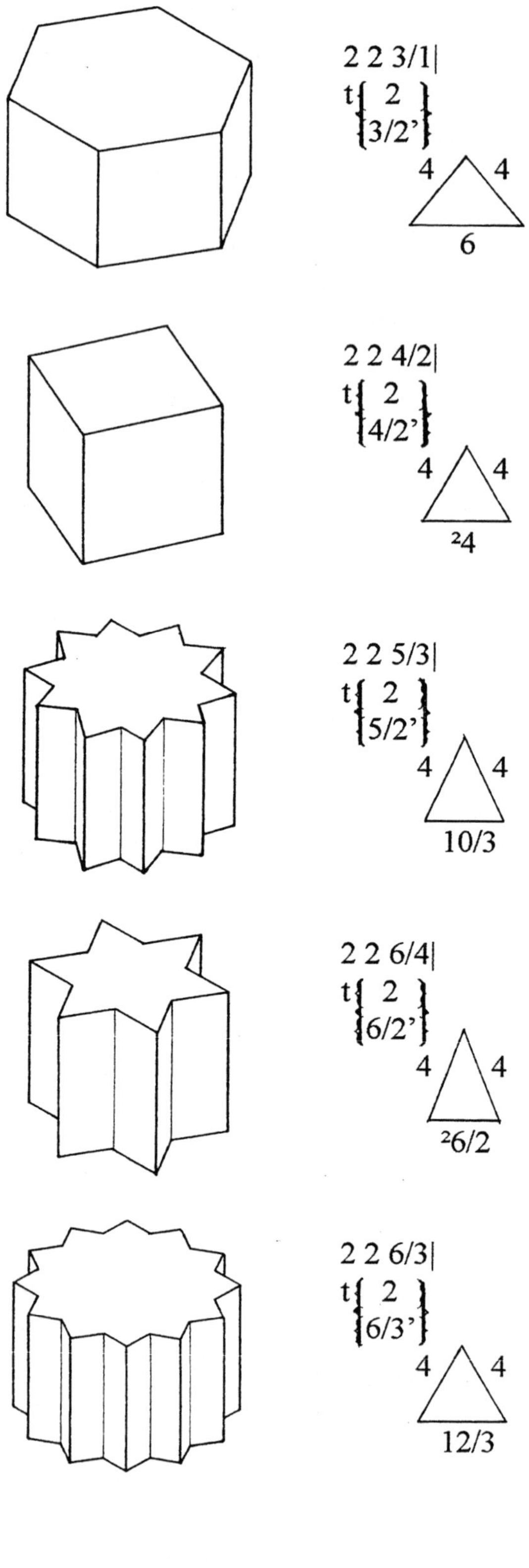

Fig.69: Quasi Even Faced Star Dihedra

SUMMARY

Except for the lack of a final super-family, the polyhedra we have examined in this volume have run exactly in parallel with the tilings we observed in the second half of the earlier volume '*Incomplete Tilings*'.

We started with a look at two super-families (3 4 4/3) and (3 5 5/3), formed by pairing up the regular polyhedra sub-family Schwarz triangles. We then continued by examining the three symmetry families of cross and star polyhedra generated by three further Schwarz triangles (4 2 4/2), (5 2 5/2) and (3 2 5/2). These included simple truncated forms, snubs and the more complex quasitruncated forms for each family.

Further Schwarz triangles, consisting of pairs of these Schwarz triangles back to back, were then examined leading to further star polyhedral forms in five sub-families. The generating triangles of the two super-families, three star families and their five related sub-families, essentially the more complex polyhedra, are shown in Figure 70, along with a typical generating triangle for the star dihedral families (2 n/d 2).

Coxeter et al. in their famous 1953 paper '*Uniform Polyhedra*' include members from all these families except two, the (4 2 4/2) and (4/2 4/2 4/2). Because {4/2} is both compound and has zero area, it was not considered a legitimate polygon for inclusion. However the inclusion in the present volume of such compound faces, along with the double polygons resulting from the truncation of even denominator polygons, has hopefully now been justified by the results obtained. This has effectively filled the holes in the standard classification that previously remained curiously empty to give a more complete? picture.

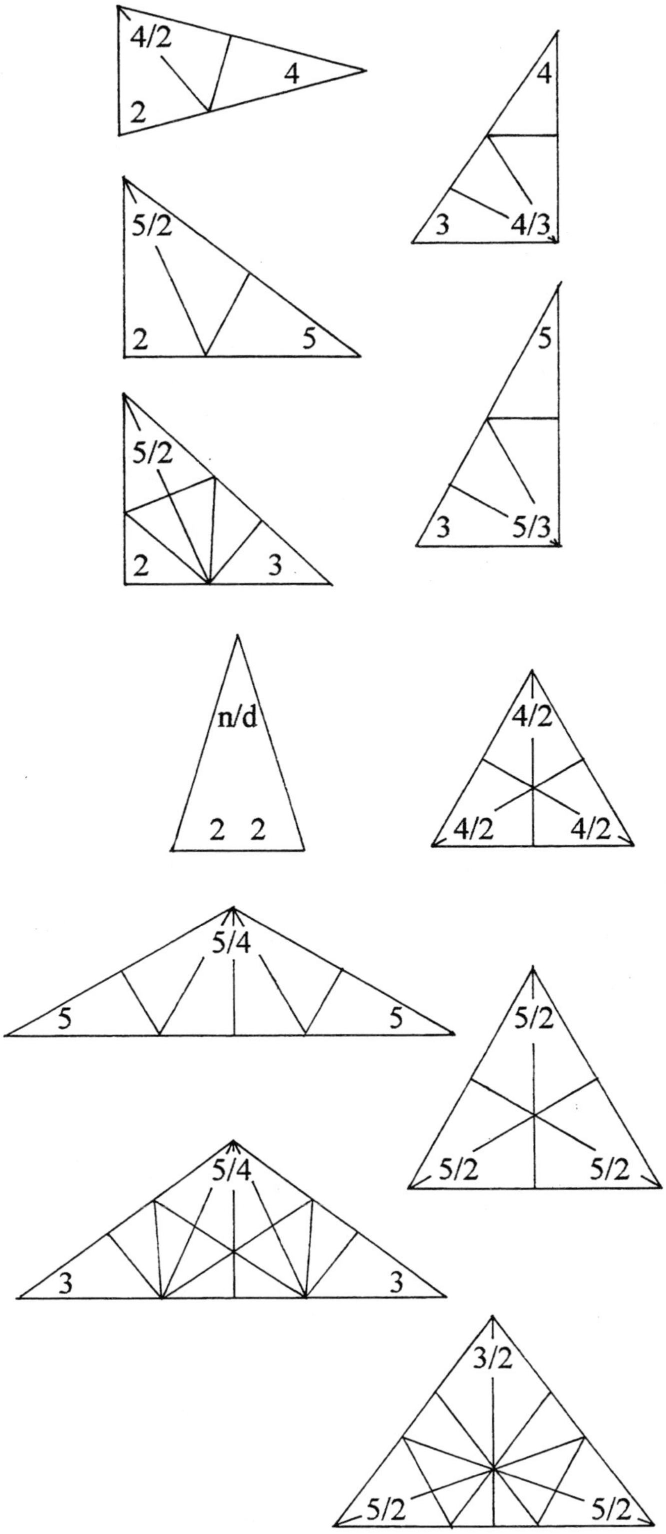

Fig. 70: Generating Triangles

SO WHAT FOLLOWS?

We have now completed our family by family review of the polyhedra, having examined all possible Schwarz triangles in the polyhedral kaleidoscopes. We have also encountered the 'Monster' polyhedra, which do not quite fit this picture, and lead one to ask whether there are any more?

According to Coxeter's classification there are not, and yet Skilling found just the one. In this volume we have encountered the 36 remaining polyhedra, acknowledged by Coxeter and Skilling as 'uniform', that were not included in the last volume. We have also seen a good many more that perhaps deserve that title.

Many of these previously disregarded forms do indeed consist of nothing more than compound forms inscribed or superimposed, however we have also seen several that can only be disregarded on account of their containing compound or double polygon faces. It remains the contention of this series that compound or double faces are as valid as any others used to construct uniform assemblies of regular polygons, and that such assemblies are equally valid as 'uniform' tilings or polyhedra.

In the next volume of this series we will continue on down Figure 71, assembling the polyhedra themselves in aggregate, firstly as closepacks filling three dimensional space, and then by having fewer or lesser valued polyhedra around an edge, we will fold the closepacks up into four dimensions to produce the hypersolids!

Further ahead still, we have the prospect of examining the skew polyhedra, assemblies of regular polygons with three dimensional vertex figures.

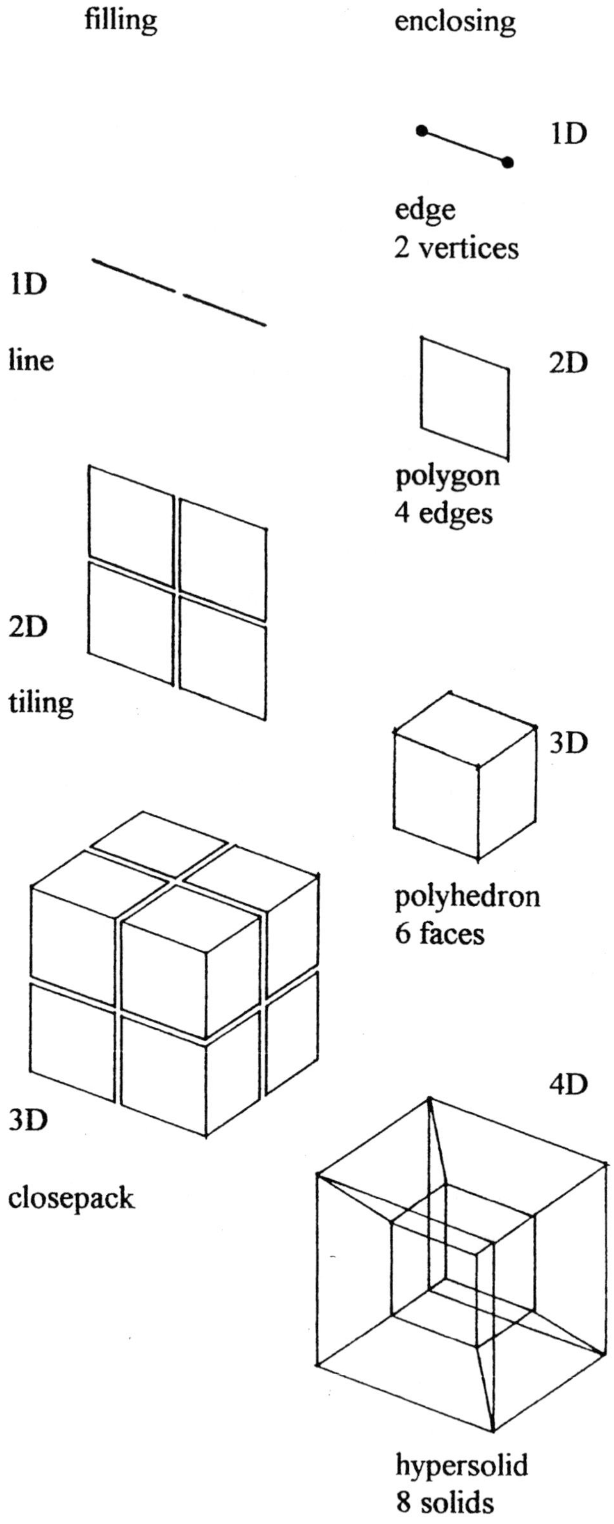

Fig.71: So What Follows?

BIBLIOGRAPHY

Abbott, E A 1962
Flatland
Blackwell

Coxeter, Longuet-Higgins & Miller 1953
Uniform Polyhedra
Phil. Trans. R. Soc. Lond. A246, p401

Coxeter, H S M 1963
Regular Polytopes
Macmillan

Critchlow, K 1969
Order in Space
Thames & Hudson

Cromwell, P R 1997
Polyhedra
Cambridge UP

Cundy, H M & Rollett, A P 1961
Mathematical Models
Oxford

Hargittai, I & M 1994
Symmetry
Shelter

Holden, A 1971
Shapes, Space and Symmetry
Columbia UP

Lakatos, I 1976
Proofs and Refutations
Cambridge UP

Peck, C E 1995
A Taxonomy of Fundamental Polyhedra
Winkworks

Skilling, J 1974
The Complete Set of Uniform Polyhedra
Phil. Trans. R. Soc. Lond. A278, p111

Taylor, P 1990
The Geometry Trip
Nattygrafix

Taylor, P 1995
Additions to the Uniform Polyhedra
Nattygrafix

Taylor, P 1997
The Complete? Polygon
Nattygrafix

Taylor, P 1998
Incomplete Tilings
Nattygrafix

Taylor, P 1999
The Simpler? Polyhedra
Nattygrafix

Wenninger, M J 1971
Polyhedron Models
Cambridge UP

Weyl, H 1952
Symmetry
Princeton UP

Williams, R 1972
Natural Structure
Constable

ALSO PUBLISHED BY NATTYGRAFIX:

THE COMPLETE? POLYGON
by Patrick Taylor 1997

being the first part of several comprising
the complete? polyhedra

presents the polygons as building blocks for the tilings, polyhedra etc.; including their generation from a simple vector model along with their truncations, quasitruncations and inverses; regular, star, compound, cross and double polygons; stellations and inscriptions.

60pp A4 paperback including
28 line drawings and 21 tables

ISBN 0 9516701 2 3 £6.00

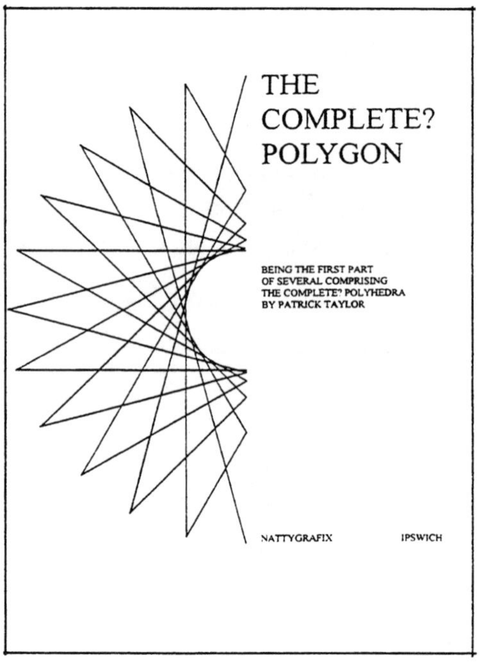

INCOMPLETE TILINGS
by Patrick Taylor 1998

being the second part of several comprising
the complete? polyhedra

presents the tilings as assemblies of polygons that fill the plane; includes truncations, snub tilings, quasitruncations; investigating in total four uniform and seven compound symmetry families using regular, star, cross, compound and double polygons

76pp A4 paperback including
80 line drawings

ISBN 0 9516701 3 1 £6.00